OPTIMALITY IN PARAMETRIC SYSTEMS

OPTIMALITY IN PARAMETRIC SYSTEMS

Thomas L. Vincent
Aerospace and Mechanical Engineering
University of Arizona

Walter J. Grantham
Mechanical Engineering
Washington State University

A WILEY-INTERSCIENCE PUBLICATION

JOHN WILEY & SONS New York • Chichester • Brisbane • Toronto

Library of Congress Cataloging in Publication Data:

Vincent, Thomas L.
Optimality in parametric systems.

Bibliography: p.
Includes index.
1. Mathematical optimization. I. Grantham, Walter J. (Walter Jervis), 1944- . II. Title.
QA402.5.V54 519 81-1870
ISBN 0-471-08307-0 AACR2

Printed in the United States of America

10 9 8 7 6 5 4 3 2 1

To Peggie Jo, Tania, Tyrone
and Tricia, Cynthia, Brian, Michelle

PREFACE

Optimization and optimal processes are integral parts of science and mathematics. Indeed, inquiries into optimal processes and optimization techniques span the history of science and mathematics. In the past the concept of "best" was often used to "explain" the rationality of nature. We quote from the Dialogues of Plato, Phaedo, 97, ca. 400 BC: "... If anyone desired to find out the course of the generation or destruction or existence of anything, he must find out what state of being or doing or suffering was best for that thing ..." Euler, in his *Methodus Inveniendi Lineas Curvas Maximi Minimiue Proprietate Gandentes*, 1744, states, "... Nothing can be met with in the world in which some maximal or minimal property is not displayed. There is, consequently, no doubt but that all the effects of the world can be derived by the methods of maxima and minima ..." Optimal concepts and processes are still a motivating force in many areas of science and are involved in such things as maximum fitness, least action, minimum potential, shortest path, and minimum cost.

Optimization is a part of mathematics because of the inherently interesting questions that may be posed, such as the problem, considered by Euclid, of finding the shortest line that may be drawn from a point to a line. As put quite simply by Maclaurin in his *Treatise of Fluxions*, Vol. 1, 1742, "There are hardly any speculations in geometry more useful or more interesting than those which relate to maxima and minima." Optimization techniques, as used in modern engineering design, illustrate the practical side as well.

This is a book on optimality concepts and optimization techniques, including multiple-objective systems (i.e., games). In studying optimization techniques, it is natural to divide the subject not only according to the optimality concept involved (min, min-max, Nash, Pareto, etc.), but also according to the mathematics required to obtain a solution. Roughly speaking, problems involving the optimal selection of *parameters* may be solved using the differential calculus, whereas problems involving the optimal selection of *functions* require the calculus of variations (or optimal control theory) for a solution. As the title of this text indicates, we are concerned with optimality in parametric systems. The optimality concepts for multiple-objective parametric systems, however, also apply directly to problems in the calculus of variations, optimal control theory, and dynamic game theory. By concentrating on parametric systems, we are able to focus on the fundamental optimality concepts without additional complexities.

This text is devoted to a unified theoretical approach to parameter optimization, encompassing nonlinear static and dynamic systems with multiple objectives. The topics include linear and nonlinear programming, vector-valued costs, continuous games, and parametric dynamical systems. Compared with optimal control theory, parameter optimization has not received as much theoretical attention in textbooks such as this in recent years. The available textbooks on game theory, for example, are generally restricted to matrix games, linear programming, duality, and mixed strategy solutions. In contrast, this text considers parametric game theory in the setting of a general nonlinear programming system with equality and inequality constraints, but without recourse to mixed strategies. In addition to the nonlinear programming setting, this text also introduces state and control notation and considers parameter optimization in dynamic systems. Thus the text not only develops a unified theory of optimality in parametric systems, but also provides an introduction to the types of systems considered in optimal control theory. Furthermore, using state and control notation, certain misconceptions are clarified concerning Lagrange multipliers and maximum principles. Much of the material in this text is not currently available in textbook form. Some of the material has never before been published, such as certain solution concepts and the material on parameter optimization in dynamic systems.

This text is not directly concerned with numerical solution techniques. For some of the optimality concepts considered in this text, numerical techniques have not yet been developed. Where such techniques do exist there are many excellent texts devoted entirely to numerical methods. The usefulness of a theoretical text in an age when many practitioners depend on a digital computer to "optimize" for them lies in the simple fact that an understanding of optimal processes is often necessary for interpreting results, whether they are obtained numerically or analytically. Furthermore, readers will find that the necessary conditions and sufficient conditions developed in the text, along with the procedures illustrated in the examples, frequently will lead either to analytical solutions or to simplified numerical solutions to many difficult practical problems.

This text is suitable for a one-semester introductory course in optimization and game theory for senior or first-year graduate students in engineering, mathematics, economics, and other areas, such as mathematical biology.

The student need not have a background in optimal control theory, despite the fact that the text contains a few comparisons with optimal control theory. The principal mathematical tool is Taylor's theorem, although certain geometric topics in linear algebra (cones and separating hyperplanes) are also employed. With the exception of differential equations, employed in the last chapter, the required mathematical background is presented in the beginning of the text.

Chapter 1 discusses the class of problems contained in this text and provides some explanation of why a text on parametric optimization involves more than setting a derivative equal to zero. The latter half of the chapter contains mathematics required for the rest of the text. Readers may wish to familiarize themselves with what is available here and then refer back to Chapter 1 as required in later chapters.

Chapter 2 contains a modern approach to parametric optimization using tangent vectors and tangent cones for the development of both first- and second-order conditions for the basic nonlinear programming problem. We feel that a thorough understanding of this approach to parametric optimization lays a solid foundation that is not only useful in the understanding and implementation of numerical methods available elsewhere, but is essential for a host of new problems in parametric optimization not usually envisioned by the student. These include vector minimization, continuous games, and parametric dynamic systems.

The objective of this book is to work from the theoretical base of Chapter 2 into many of the new concepts of parameter optimization. These concepts are currently used to some extent in biology, economics, and engineering. It is our belief that, in the future, they will be much more generally applied as theoretical results become more widely known. Many examples are included in the text to illustrate the applicability of these concepts to various disciplines.

A fairly comprehensive treatment of vector-valued cost criteria is given in Chapter 3. This treatment draws on the methods of Chapter 2 and relies quite heavily on current theoretical work in this area.

Chapter 4 is a transition chapter, which reformulates the nonlinear programming problem in terms of state and control notation. This notation is needed for the last two chapters. Here is a rigorous translation from the classical notation usually associated with parametric optimization to the useful control notation favored by many engineers. The first- and second-order results of Chapter 2 are again obtained in control notation and the computational advantage of this approach is illustrated. The Lagrange multipliers may now be interpreted in terms of sensitivity coefficients and their role with respect to constrained extrema are examined and clarified through the use of examples.

The notation of Chapter 4 paves the way for using the methods of Chapter 2 in a simple yet comprehensive treatment of continuous static games. In Chapter 5, solution concepts for games are introduced and a theory is developed that allows for the solution of numerous examples and exercises given in this chapter.

Chapter 6 applies the methods of the previous chapters to parametric dynamic systems, that is, systems in which the time histories of the state variables are governed by ordinary differential equations, where the control inputs are either constants or specified functions containing control parameters. In general, the differential equations defining the system are not integrable. Therefore the necessary conditions developed in this chapter involve an auxiliary system of differential equations similar but not equivalent to the adjoint system used with the maximum principle in optimal control theory. Several examples are also included in this chapter to illustrate the methods and applicability of the theory.

In the exercises given at the end of each chapter we have attempted to show the types of questions or problems that may be formulated in a number of different fields. We feel that the surface has barely been scratched in this regard. One objective of this text is to provide an awareness of the types of parametric optimization questions that may be asked in a game theoretic setting.

THOMAS L. VINCENT
WALTER J. GRANTHAM

Tucson, Arizona
Pullman, Washington
May 1981

CONTENTS

SYMBOLS AND NOTATION

$=$	Equals
$\neq$	Does not equal
$\triangleq$	Equals by definition
$\equiv$	Equals identically
$\geqq$ $(\leqq)$	Greater (less) than or equal to
$\geqslant$ $(\leqslant)$	Partially greater (less) than (See Section 3.2)
$>$ $(<)$	Greater (less) than
$\forall$	For all
$\in$	Is an element of
$\notin$	Is not an element of
$\cup$	Union
$\cap$	Intersection
$\subset$	Is a subset of
$\subseteq$	Is a subset or all of
$A-B$	Set of elements in A but not in B
$\varnothing$	Empty set
$[a, b]$	Closed interval $a \leqslant y \leqslant b$
(a, b)	Open interval $a<y<b$
$(a, b]$	Semiopen interval $a<y \leqslant b$
E^m	Euclidean space of dimension m
$\|\cdot\|$	Euclidean norm
$\vert\cdot\vert$	Determinate, absolute value
$\{A \vert B\}$	The set of all A such that B holds
$G(\cdot): A \rightarrow B$	G maps A into B
$\mathring{Y}$	Interior of set Y
$\bar{Y}$	Closure of set Y
∂Y	Boundary of set Y

OPTIMALITY IN PARAMETRIC SYSTEMS

Chapter One

FUNDAMENTALS

1.1 INTRODUCTION

There are a wide variety of system design problems, found in most fields of scientific endeavor, that require the determination of parameters, under imposed constraints, such that a given specification of system performance is optimized. We use the terminology *parametric systems* to designate the class of systems for which such an optimal process can be realized.

The parametric systems examined in this text include both static and dynamic systems. The primary emphasis is on static systems whose parameters are constrained by a system of algebraic equality and inequality equations. The dynamic systems, introduced in Chapter 6, are control systems whose state is governed by ordinary differential equations and whose control variables are parameters (i.e., constants to be selected).

We begin our discussion of parametric optimization with static systems that may be characterized by a *cost* or *costs* to be minimized by a choice of a finite number of parameters $y_1, y_2, \ldots, y_m$. This choice may be thought of as selecting a point (i.e., a vector) $y=(y_1, y_2, \ldots, y_m)$ in an m-dimensional Euclidean space E^m. The system is static in the sense that no time history is involved for the choice y. No loss of generality is incurred by concentrating on minimization problems, since maximizing $H(y)$ can be accomplished by minimizing $G(y)=-H(y)$.

For most problems of practical interest the choice for y cannot be made over the entire Euclidean space E^m. In particular, we are interested in systems in which the choice for y is restricted by n scalar-valued *equality constraint* functions of the form

$$\begin{aligned} g_1(y)&=0 \\ &\vdots \\ g_n(y)&=0. \end{aligned} \tag{1.1}$$

In general, $n<m$; however, this is not a required restriction unless the

equality constraints are independent. If $m=n$ and the constraints are independent, then (1.1) has a locally unique solution and no optimization is possible. If $n>m$, then either some of the constraints are redundant (which is allowable) or, if the constraints are truly independent, then (1.1) has no feasible solution.

In addition to the equality constraints (1.1), the choice for y may also be restricted by q scalar-valued *inequality constraint* functions of the form

$$\begin{aligned} &h_1(y)\geqq 0\\ &\vdots\\ &h_q(y)\geqq 0. \end{aligned} \tag{1.2}$$

For convenience we use a shorthand notation for (1.1) and (1.2). Letting $N=\{1\cdots n\}$ and $Q=\{1\cdots q\}$, we may rewrite (1.1) and (1.2) as

$$g_i(y)=0 \qquad \forall i\in N \tag{1.3}$$

$$h_j(y)\geqq 0 \qquad \forall j\in Q. \tag{1.4}$$

Employing this subscript notation, the set of points available for computing costs, that is, the *constraint set*, is given by

$$Y=\{y\in E^m \mid g_i(y)=0\ \forall i\in N \text{ and } h_j(y)\geqq 0\ \forall j\in Q\}. \tag{1.5}$$

Note that, if N and Q are empty, then $Y=E^m$.

The expressions may be shortened further through the use of vector notation. Let y, $g(y)$, and $h(y)$ be represented by the following column vectors:

$$\begin{aligned} y&=[y_1,\ldots,y_m]^T\\ g(y)&=[g_1(y),\ldots,g_n(y)]^T\\ h(y)&=[h_1(y),\ldots,h_q(y)]^T. \end{aligned}$$

Then (1.3)–(1.5) may be written in vector form as

$$g(y)=0 \tag{1.6}$$

$$h(y)\geqq 0 \tag{1.7}$$

$$Y=\{y\in E^m \mid g(y)=0 \text{ and } h(y)\geqq 0\} \tag{1.8}$$

where† $g(\cdot)$: $E^m \to E^n$, $h(\cdot)$: $E^m \to E^q$, and $Y \subseteq E^m$. The notations $=$ and $\geqq$ apply to each component of $g(y)$ and $h(y)$, respectively, and $[\ ]^T$ denotes the transpose of $[\ ]$.

For each choice of $y \in Y$ one or more costs may be defined by means of a set of scalar-valued functions $G_1(y), G_2(y), \ldots, G_r(y)$. For the case where $r=1$ we have a *scalar-valued cost criterion*. For the case $r>1$ we have either a *vector-valued cost criterion* (i.e., a "trade-off analysis" problem) or a *continuous static game*. In order to distinguish between the latter two cases, additional formulation is necessary; in particular, we must establish whether one or more *players* are involved in selecting the point $y \in Y$ and the relation of the various costs to the various players.

Chapters 2 and 4 deal with the single-criterion/single-player case where $r=1$, and Chapters 3, 5, and 6 deal with the case where $r>1$. The additional formulation needed for the multicriteria case is postponed until Chapter 3.

1.2 THE SCALAR COST PROBLEM

In dealing with the single-criterion case $r=1$, the subscript on G_1 is dropped. Only in this case is it possible to define a minimum in the usual sense. For this case we say that a point $y^* \in Y$ is *optimal* if $G(\cdot)$ takes on a *global minimum* value, that is,

$$G(y^*) \leqq G(y)$$

for all $y \in Y$. If $G(y^*) < G(y)$ for all $y \in Y$, $y \neq y^*$, then $G(\cdot)$ is said to take on a proper minimum at y^*. More than one point in Y may be optimal. However, it follows from the above definition that the minimum cost is unique.

We are interested in developing a number of constructive conditions that are useful in determining solutions to various optimization problems. To do this certain assumptions must be made regarding the functions $G(\cdot)$, $g_i(\cdot)$, and $h_j(\cdot)$. So far, use of the functional notation implies that the mapping of points from Y to E^1 is defined for each of the functions for every point of Y and that a given point of Y is mapped to a unique point of E^1. The following paragraphs provide motivation for the additional assumptions to be made.

Consider the problem of minimizing a scalar cost whose value is determined by the choice of a single parameter (i.e., $y \in E^1$). We examine several cases with the same cost to illustrate the significance of the various assumptions we make.

† The expressions $g(y)$, $g(\cdot)$, and g are all used to denote the function or its value.

Example 1.1 (Cost Undefined) We take our cost to be

$$\text{cost} = \frac{1}{y}$$

where $y \in Y \subset E^1$. We impose no equality constraints ($n=0$). We do, however, impose two ($q=2$) inequality constraints on the choice of y:

$$h_1(y) = 5 - y \geqq 0$$

$$h_2(y) = y + 2 \geqq 0.$$

Thus the constraint set Y is given by

$$Y = \{ y \in E^1 | -2 \leqq y \leqq 5 \}.$$

It is helpful to plot the cost for various values of y as shown in Figure 1.1. The cost is undefined at $y=0$, and the cost has no minimum in this case. In fact, there are no bounds on the cost. Note that, since the cost was undefined at $y=0$, it is improper to use the notation $G(y)$ to designate the cost for all $y \in Y$.

Example 1.2 (Greatest Lower Bound) Again, we take the scalar cost as

$$G(y) = \frac{1}{y}$$

where $y \in Y \subset E^1$. Again, we employ no equality constraints ($n=0$). The

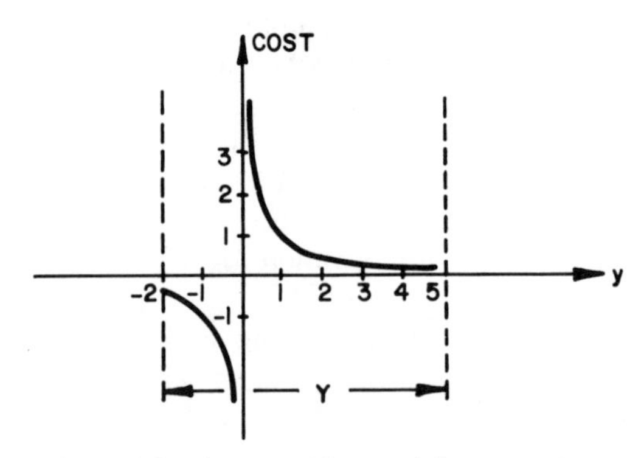

Figure 1.1. A case with no minimum cost.

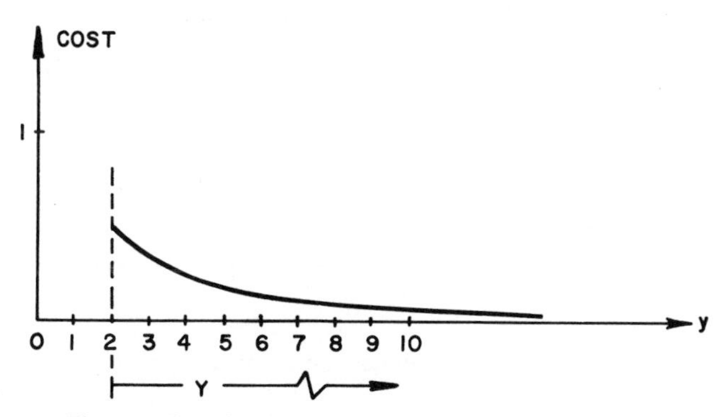

Figure 1.2. No minimum cost, lower bound on cost.

single inequality constraint is

$$h_1(y) = y - 2 \geqq 0,$$

so the constraint set Y is given by

$$Y = \{y \in E^1 | y \geqq 2\}.$$

Note that from every point of Y the mapping $G(\cdot)$ is defined and unique so that we can use the notation $G(y)$ for the cost. The graph of this function is illustrated in Figure 1.2.

Again, there is no minimum cost in this case since $G(y)$ decreases to zero asymptotically as $y \to \infty$ (the reader may test this statement using the definition of a global minimum). Lower bounds to the cost exist, however. For example, the cost is always greater than -10. In fact, there is a *greatest lower bound* ("glb," "inf," "infimus"). If a set fails to have a minimum, the greatest lower bound is often used to take its place. In this case $\inf(G) = 0$.

Example 1.3 (Constrained Minimum) We retain the cost

$$G(y) = \frac{1}{y}$$

where $y \in Y \subset E^1$. Again, no equality constraints are imposed ($n = 0$). The constraint set Y is defined by the inequality constraints

$$h_1(y) = y - 2 \geqq 0$$

$$h_2(y) = 5 - y \geqq 0,$$

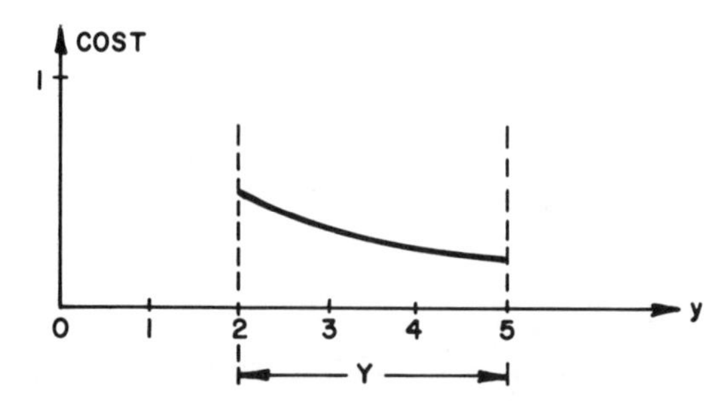

Figure 1.3. Minimum cost on the boundary of Y.

that is,

$$Y=\{y\in E^1|2\leqq y\leqq 5\}.$$

The graph of the cost $G(y)$ is shown in Figure 1.3. In this example the minimum cost does exist and it occurs on the *boundary* of the set Y.

Example 1.4 (Multiple Minima) In this case we take the cost as

$$G(y)=\begin{cases}3 & \text{if } y\leqq 3\\ 2 & \text{if } y>3\end{cases}$$

with $y\in E^1$ and no other constraints imposed ($n=q=0$). The graph of the cost is shown in Figure 1.4. In this example the minimum cost is 2, and it occurs for all values of $y>3$. The minimum cost is unique, but more than one point is optimal.

It is clear that, without certain restrictions on the cost relation and on the set Y, we have no guarantee of a minimum cost. The restrictions

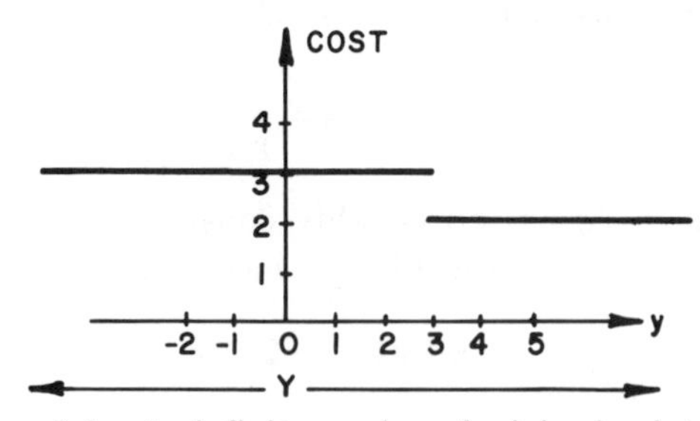

Figure 1.4. An infinite number of minimal points y^*.

sufficient to guarantee a minimum are conveniently given to us by the *theorem of Weierstrass*. According to this theorem (see Section 1.4), for a minimum cost to occur at some point in Y, it is sufficient that the cost be representable by a function whose value is defined at each point $y \in Y$ (rules out Example 1.1), that the cost be continuous (rules out Example 1.4), and that the set Y be compact (closed and bounded—rules out Example 1.2). Note that the cost may have a minimum even though the sufficiency conditions for the theorem of Weierstrass are violated. The theorem simply states conditions under which we are guaranteed a minimum. Of the above four examples only Example 1.3 satisfies all of the requirements.

Example 1.5 Determine the global minimum for the function $G(y) = 2y^3 - 3y^2 - 36y + 3$ where $y \in Y = \{y \in E^1 | y + 5 \geqq 0 \text{ and } 4 - y \geqq 0\}$. In this case the theorem of Weierstrass is satisfied. The global minimum is easily found to be at $y = -5$, with $G(-5) = -142$, by simply examining the plot of $G(y)$ over the range $-5 \leqq y \leqq 4$. Note that the slope of this function is zero at $y = -2$ and $y = 3$, with $d^2G/dy^2 > 0$ at $y = 3$. Both the points $y = 3$ and $y = -5$ correspond to *local* minimum points for $G(\cdot)$ on Y.

It is apparent that, if we are seeking a minimum for a continuous cost function over a compact set, the minimum may occur at points either where all inequality constraints are satisfied as strict inequalities or where one or more inequality conditions are satisfied as equalities. This point is significant in the development of necessary conditions and sufficient conditions used to determine minimal points.

In what follows we obtain a number of necessary conditions and a number of sufficient conditions for a point $y^* \in Y$ to be a minimal point. A given necessary condition and a given sufficient condition taken together are both necessary and sufficient if and only if the restrictions used for both conditions are satisfied. As a consequence, we adopt the following approach:

1. The necessary conditions are used to select suitable points $y \in Y$ that are candidates for a minimum.
2. These points are then checked by use of sufficient conditions.
3. If a sufficient condition is satisfied, then the point in question yields a minimum value to the function.
4. If a sufficient condition is not satisfied, the point in question still remains as a candidate for a minimum.

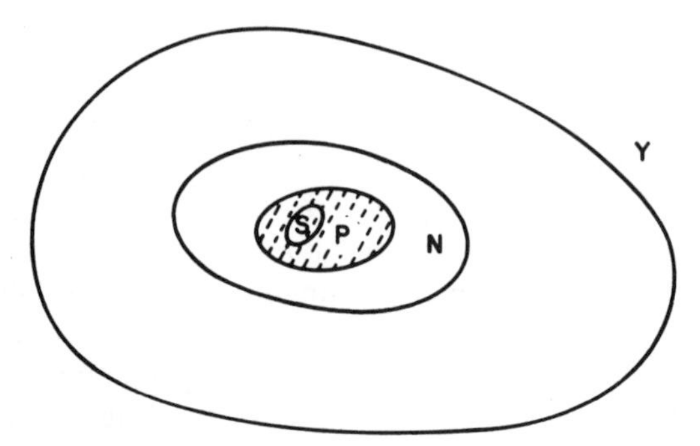

Figure 1.5. Relationship of minimal points to those that satisfy necessary conditions and sufficient conditions.

To illustrate the relation between minimal points and necessary and sufficient conditions, let P be the set of actual minimal points for a cost function whose domain is Y. Let N be the set of points that satisfy some necessary conditions for a minimum cost, and let S be the set of points that satisfy *any* sufficient conditions for a minimum cost. Figure 1.5 illustrates the fact that $S \subseteq P \subseteq N$, provided that the cost function and the domain Y satisfy all restrictions imposed in obtaining the necessary conditions and sufficient conditions. Less precisely, satisfaction of all necessary conditions does not guarantee a minimum. However, use of one or more necessary conditions narrows the field of candidates. On the other hand, satisfaction of even a single sufficiency condition guarantees a minimum. For this reason, sufficiency conditions are generally more restrictive than necessary conditions.

1.3 DEFINITIONS

Before proceeding we introduce some definitions and conventions that we employ throughout this text.

1.3.1 Fundamental Terms

Set A collection of objects that have common and distinguishing properties.

Function A correspondence or mapping between two sets A and B that pairs each member $a \in A$ with a unique member $\phi(a) = b \in B$. The set A is called the domain of definition of $\phi(\cdot)$ and the set B is called the range of values of $\phi(\cdot)$. In symbols we write: $\phi(\cdot): A \rightarrow B$.

Class Let y be a point in $Y \subseteq E^m$. Then a function $\phi(\cdot): Y \rightarrow E^1$ that is continuous on Y is said to be of class C^1 on Y if the first-order partial derivatives $\partial\phi/\partial y_i$, $i=1,\ldots,m$, are defined and continuous on Y. It is said to be of class C^2 if it is of class C^1 and the second-order partial derivatives

$$\frac{\partial^2\phi}{\partial y_i \partial y_j}$$

are defined and continuous on Y, where $i, j=1,\ldots,m$.

Differentiation with Respect to a Vector By convention all vectors are column vectors unless otherwise specified. If $\phi(y)$ is a C^1 scalar-valued function of a vector $y \in E^m$, that is, if $\phi(\cdot): E^m \rightarrow E^1$, then

$$\frac{\partial\phi}{\partial y} \triangleq \left[\frac{\partial\phi}{\partial y_1}, \ldots, \frac{\partial\phi}{\partial y_m}\right] \tag{1.9}$$

is defined as a *row* vector with m components. Similarly, if $f(y)$ is a (column) vector in E^s representing a C^1 vector-valued function $f(\cdot): E^m \rightarrow E^s$, then

$$\frac{\partial f}{\partial y} \triangleq \begin{bmatrix} \frac{\partial f_1}{\partial y_1} & \cdots & \frac{\partial f_1}{\partial y_m} \\ \vdots & & \\ \frac{\partial f_s}{\partial y_1} & \cdots & \frac{\partial f_s}{\partial y_m} \end{bmatrix} \tag{1.10}$$

is an $s \times m$ matrix whose jth row is $\partial f_j/\partial y$.

The convention adopted in this text is that the operation $\partial(\ )/\partial y$ produces a row vector. If this operation is to be applied to a row vector, that is, if $f^T=[f_1,\ldots,f_s]$, the row vector is first transposed and then the operation is applied to the resulting column vector:

$$\frac{\partial}{\partial y}(f^T) \triangleq \frac{\partial f}{\partial y}. \tag{1.11}$$

The following are some useful formulas for differentiation with respect to vectors. Let

$$x=[x_1,\ldots, x_s]^T$$

$$y=[y_1,\ldots, y_m]^T$$

$$g(y)=[g_1(y),\ldots, g_n(y)]^T$$

$$f(y)=[f_1(y),\ldots, f_n(y)]^T,$$

and let A and B be $m\times m$ and $s\times m$ constant matrices, respectively. Then

$$\frac{\partial}{\partial y}[f+g]=\frac{\partial f}{\partial y}+\frac{\partial g}{\partial y}$$

$$\frac{\partial}{\partial y}[f^Tg]=f^T\frac{\partial g}{\partial y}+g^T\frac{\partial f}{\partial y}$$

$$\frac{\partial}{\partial y}[x^TBy]=x^TB$$

$$\frac{\partial}{\partial x}[x^TBy]=\frac{\partial}{\partial x}[y^TB^Tx]=y^TB^T$$

$$\frac{\partial}{\partial y}[y^TAy]=y^T\frac{\partial}{\partial y}(Ay)+y^TA^T\frac{\partial}{\partial y}(y)$$

$$=y^TA+y^TA^T$$

$$=2y^TA \quad \text{if} \quad A=A^T$$

$$\frac{\partial^2}{\partial y^2}[y^TAy]=A+A^T$$

$$=2A \quad \text{if} \quad A=A^T.$$

Example 1.6 Let $y=[y_1, y_2]^T$ be a vector in E^2 and let $\phi(\cdot)$: $E^2\to E^1$ be the scalar-valued function

$$\phi(y)=y_1^2\sin(y_2)+2y_1y_2.$$

Then $\partial\phi/\partial y$ is the 1×2 row vector

$$\frac{\partial\phi}{\partial y}=\left[\frac{\partial\phi}{\partial y_1},\frac{\partial\phi}{\partial y_2}\right]$$

$$=\left[2y_1\sin(y_2)+2y_2,\ y_1^2\cos(y_2)+2y_1\right],$$

and $\partial^2\phi/\partial y^2$ is the 2×2 matrix

$$\frac{\partial^2\phi}{\partial y^2}\triangleq\frac{\partial}{\partial y}\left[\frac{\partial\phi}{\partial y}\right]^T$$

$$=\begin{bmatrix}\dfrac{\partial}{\partial y}(2y_1\sin(y_2)+2y_2)\\ \dfrac{\partial}{\partial y}(y_1^2\cos(y_2)+2y_1)\end{bmatrix}$$

$$=\begin{bmatrix}2\sin(y_2) & 2y_1\cos(y_2)+2\\ 2y_1\cos(y_2)+2 & -y_1^2\sin(y_2)\end{bmatrix}.$$

1.3.2 Vector Spaces and Matrices

In what follows superscripts are used to denote vectors, and subscripts denote scalars and scalar components of a vector, that is, $x=(x_1,\ldots,x_m)^T$, $y=(y_1,\ldots,y_m)^T$, $y^i=(y_1^i,\ldots,y_m^i)^T$ are all used to denote vectors. Let R be the set of real numbers.

Vector Space Let Y be a nonempty set with rules of vector addition and scalar multiplication (defined below) that assign to any vectors $y^1, y^2\in Y$ a sum $y^1+y^2\in Y$ and to any $y\in Y$ and any real number k a product $ky\in Y$, respectively. Then Y is called a vector space over the real numbers R (and the elements of Y are called vectors) if the following axioms hold:

1 $(y^1+y^2)+y^3=y^1+(y^2+y^3)$.
2 There exists an element in Y denoted by 0 such that $y+0=y$.
3 For each vector y there is a vector $\in Y$, denoted by $-y$, for which $y+(-y)=0$.
4 $y^1+y^2=y^2+y^1$.
5 $k(y^1+y^2)=ky^1+ky^2 \qquad \forall k\in R$.

6 $(a+b)y=ay+by \qquad \forall a, b \in R.$
7 $(ab)y=a(by) \qquad \forall a, b \in R.$

Note that here y, y^1, y^2, and y^3 are arbitrary elements of Y.

Euclidean Space A vector space over the field of real numbers is a Euclidean space, denoted by E^m, if it has the additional property that the "length" or "norm" of a vector $y=(y_1,\dots, y_m)^T \in E^m$ is defined as

$$\|y\| \triangleq \sqrt{y_1^2+\cdots+y_m^2} \triangleq \sqrt{y^T y}\,.$$

Vector Addition $y+x \triangleq (y_1+x_1,\dots, y_m+x_m)^T.$

Scalar Multiplication by k $ky \triangleq (ky_1,\dots, ky_m)^T.$

Dot Product $x\cdot y=x^T y \triangleq x_1 y_1+\cdots+x_m y_m.$

Subspace Let X be a subset of a vector space Y. X is called a subspace of Y if X is itself a vector space with respect to the operations of vector addition and scalar multiplication on Y. For example, E^2 is a subspace of E^3.

Span A vector space Y is said to be spanned or generated by the m vectors $y^1,\dots, y^m$ provided that each y^i is in Y and that every vector in Y can be written as a *linear combination* of the y^i, that is, every vector $y \in Y$ is of the form

$$y=k_1 y^1+\cdots+k_m y^m$$

where the k_i are real numbers.

Row and Column Space of a Matrix Let A be an arbitrary $m \times n$ matrix of real numbers. The rows of A, viewed as vectors in E^n, span a subspace of E^n called the row space of A. The columns of A, viewed as vectors in E^m, span a subspace of E^m called the column space of A.

Linear Dependence Let Y be a vector space over R. The vectors $y^1,\dots, y^m \in Y$ are said to be linearly dependent if there exist $k_1,\dots, k_m \in R$, *not all* *zero*, such that the following equation is satisfied:

$$k_1 y^1 + \cdots + k_m y^m = 0.$$

Otherwise, the vectors are said to be linearly independent.

Dimensions A vector space Y is said to be m-dimensional, written $\dim(Y) = m$, if there exist exactly m linearly independent vectors $y^1, \ldots, y^m \in Y$. The set $\{y^1, \ldots, y^m\}$ is called a *basis* of Y.

Rank of a Matrix Let A be an arbitrary $m \times n$ matrix of real numbers. The dimensions of the row space and column space of A are called, respectively, the row rank and the column rank of A. The rank of the matrix A is the common value of its row rank and column rank [see Nering (1963) or any linear algebra text for a proof of the equivalence of row and column rank]. The rank of a matrix gives the maximum number of linearly independent rows and also the maximum number of linearly independent columns.

Rank of a Matrix (Alternate Definition) A nonzero $n \times m$ matrix A is said to have rank r if at least one of its r-square minors (obtained by deleting rows and columns of A) is different from zero while every $(r+1)$ square minor, if any, is zero. A zero matrix is said to have rank zero.

Rank of a Square Matrix Let A be an m-square matrix over R. The rank of A is equal to the order of the largest square submatrix of A (obtained by deleting rows and columns of A) whose determinant is not zero.

Nonsingular Matrix An m-square matrix A is called nonsingular if its rank $r = m$, that is, if $|A| \neq 0$ where $|\cdot|$ denotes the determinant. Otherwise, A is called singular.

Example 1.7 Let A be the 2×4 matrix

$$A = \begin{bmatrix} 1 & 2 & -2 & 1 \\ 0 & 1 & -1 & -4 \end{bmatrix}.$$

The row space of A is the set of all vectors $x \in E^4$ of the form

$$x = k_1 \begin{bmatrix} 1 \\ 2 \\ -2 \\ 1 \end{bmatrix} + k_2 \begin{bmatrix} 0 \\ 1 \\ -1 \\ -4 \end{bmatrix}$$

where k_1 and k_2 are real numbers. The dimension of the row space of A is 2 (the largest number of linearly independent rows of A). The column space of A is the set of all vectors $y \in E^2$ of the form

$$y = k_1\begin{bmatrix}1\\0\end{bmatrix} + k_2\begin{bmatrix}2\\1\end{bmatrix} + k_3\begin{bmatrix}-2\\-1\end{bmatrix} + k_3\begin{bmatrix}1\\-4\end{bmatrix}$$

where the k_i are real numbers. The dimension of the column space of A is 2. Thus the rank of A is 2. This result can also be determined by examining the 2×2 determinants formed by pairs of columns of A.

1.3.3 Sets

Ball About a Point y^* Suppose $y^* \in Y \subseteq E^m$ and $\delta y \in E^m$. Then a ball about y^* is any of the sets B given by

$$B = \{ y^* + \delta y \mid \| \delta y \| < \varepsilon \text{ for any finite } \varepsilon > 0 \}.$$

Interior Point of Set Y The point y^* is interior to Y if $y^* \in Y$ and if there exists a ball B about y^* that contains only points of Y, that is, if $B \subseteq Y$. The set of all points interior to Y is denoted by $\mathring{Y}$.

Exterior Point of Set Y The point y^* is exterior to Y if $y^* \notin Y$ and if there exists a ball B about y^* that contains no point of Y, that is, if $B \cap Y = \varnothing$. The exterior of Y is denoted by comp($\bar{Y}$), where $\bar{Y}$ is the closure of Y (see below).

Boundary Point of Set Y The point y^* is a boundary point for Y if y^* is neither interior to Y nor exterior to Y.

Boundary to Set Y The set of all boundary points for Y, denoted by ∂Y.

Closed Set A set that contains all its boundary points. The closure of Y is denoted by $\bar{Y}$. Thus Y is closed if and only if $Y = \bar{Y}$.

Open Set A set that contains none of its boundary points (e.g., a ball B).

Bounded Set A set $Y \subset E^m$ is bounded if there exists a ball B such that $Y \subset B$.

Compact Set A set $Y \subset E^m$ is compact if it is both closed and bounded.

Convex Set A set Y is said to be convex if, for any $y \in Y$ and $y+\delta y \in Y$, the point $y+\alpha\delta y \in Y$ for all $\alpha \in [0,1]$.

Local Convexity The set Y is locally convex at $y^* \in Y$ if there exists a ball B with a center at y^* such that $B \cap Y$ is convex.

Arcwise Connected A set Y is arcwise connected if any two of its points can be joined by a polygonal line that lies entirely in Y.

Locally Arcwise Connected A set Y is locally arcwise connected at $y^* \in Y$ if there exists a ball B with a center at y^* such that $B \cap Y$ is arcwise connected.

Hyperplane Let $a=[a_1,\ldots,a_m]^T$ be a nonzero vector $\in E^m$ and let b be a scalar. Then a hyperplane is the set of points $y \in E^m$ such that $a^T y = b$.

Cone A nonempty set $C \in E^m$ is a cone if for every real number $\beta \geqq 0$, $y \in C$ implies $\beta y \in C$. Let $\{y^1,\ldots,y^q\}$ be a set of vectors in E^m and let a_i, $i=1,\ldots,q$, denote real numbers. The set

$$C=\left\{y \in E^m \,|\, y=\sum_{i=1}^{q} a_i y^i, a_i \geqq 0\right\}$$

is called the finite convex cone (or simply the convex cone) generated by the vectors $\{y^1,\ldots,y^q\}$.

Polar Cone Let C be a cone. The polar cone C^* to C is defined as

$$C^*=\left\{z \in E^m \,|\, z^T y \geqq 0 \ \forall y \in C\right\}.$$

If C is a convex cone generated by $\{y^1,\ldots,y^q\}$, then $C^*=\{z \in E^m \,|\, z^T y^i \geqq 0 \ \forall i=1,\ldots,q\}$.

Dual Cone Let C be a cone. The dual cone C_D to C is defined as

$$C_D=\left\{z \in E^m \,|\, z^T y \leqq 0 \ \forall y \in C\right\}.$$

If C is a convex cone generated by $\{y^1,\ldots,y^q\}$, then $C_D=\{z \in E^m \,|\, z^T y^i \leqq 0 \ \forall i=1,\ldots,q\}$. Figure 1.6 illustrates cones in E^2.

Throughout the remainder of the text B denotes a ball centered at y^*, and we use the notation that a point near $y^* \in E^m$ may be represented by

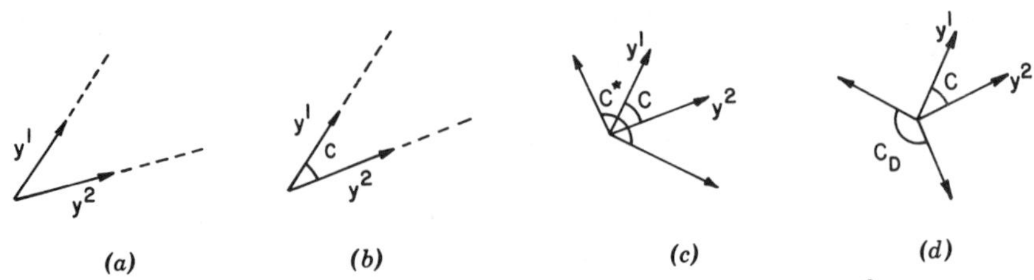

Figure 1.6. Examples of cones, polar cones, and dual cones in E^2. (a) A cone consisting of two rays: $C=\{y|y=ky^1 \text{ or } ky^2;\ k\geqq 0\}$. ($b$) A convex cone generated by y^1 and y^2: $C=\{y|y=k_1y^1+k_2y^2;\ k_1,k_2\geqq 0\}$. ($c$) Polar cone C^*. (d) Dual cone C_D.

$y^*+\delta y\in E^m$, where δy may be thought of as a small vector whose origin is at the point y^*. If $\hat{\delta y}$ is a vector in the direction of δy, then we also use the notation that $\delta y=\alpha\hat{\delta y}$, where α is a nonnegative real number.

1.4 BASIC THEOREMS

In this section we present some of the basic theorems that are employed later in the development of necessary conditions and sufficient conditions for optimality. The theorem of Weierstrass is a sufficient condition for a minimum to exist. Taylor's theorem and its corollaries provide a means for approximating functions in a neighborhood of an optimal point. Since our basic approach is to investigate the effects of small perturbations about an optimal point, Taylor's theorem plays a central role. The implicit function theorem provides sufficient conditions under which a system of equality constraints can be solved for some of the variables in terms of the other variables. The implicit function theorem forms the basis for the state and control variable notation of Chapter 4. This notation is employed again in Chapters 5 and 6 with static and dynamic games.

A consequence of employing Taylor's theorem to examine perturbations is the geometric interpretation available. We emphasize a geometric approach in developing several optimality conditions. In particular, we employ constraint cones as sets of allowable perturbation directions. Farkas' lemma provides a means of investigating whether or not a given vector is in a specified cone. Similarly, the theorem of the alternative (also known as the separating hyperplane theorem) provides a means of investigating whether or not solutions exist for given systems of equality and inequality conditions. We are concerned later with whether or not the cone of allowable perturbation directions contains a direction in which the function to be minimized is decreasing. Such points cannot be minimal points.

Theorem 1.1 (Theorem of Weierstrass) Every function $G(\cdot): Y \to E^1$ that is continuous on a compact set $Y \subset E^m$ possesses a minimum value on the interior or boundary of the set Y.

Proof The set of real numbers $\{\alpha \in R \mid \alpha = G(y), y \in Y\}$ is compact and every compact set of real numbers contains its greatest lower bound. ■

Theorem 1.2 (Taylor's Theorem) If $y^* \in E^m$ and if there exists a ball B centered at y^* such that the function $G(\cdot): B \to E^1$ is C^{r+1} for every point $y^* + \delta y \in B$, then with $\delta y = \alpha \hat{\delta y}$, where $\alpha \geqq 0$ and $\hat{\delta y}$ is a unit vector in the direction of δy, we have

$$G(y^* + \alpha \hat{\delta y}) = G(y^*) + U[G(y^*)] + \frac{1}{2!} U^2[G(y^*)] + \cdots + \frac{1}{r!} U^r[G(y^*)] + R \tag{1.12}$$

where

$$R = \frac{1}{(r+1)!} U^{r+1}[G(y^* + \beta \hat{\delta y})]$$

for some $\beta \in (0, \alpha)$ and where U is the differential operator defined by

$$U(\cdot) \triangleq \delta y_1 \frac{\partial(\cdot)}{\partial y_1} + \cdots + \delta y_m \frac{\partial(\cdot)}{\partial y_m}.$$

Proof Taylor's theorem, as presented here for functions of a vector, is a generalization of Taylor's theorem for a function of a single variable. The single-variable form of Taylor's theorem is used, along with a special single-variable function, to establish the multivariable extension of Taylor's theorem.

We wish to determine an expression for the value of $G(\cdot)$, evaluated at $y^* + \delta y$, in terms of $G(\cdot)$ and its derivatives evaluated at y^*. We begin by parameterizing the point at which $G(\cdot)$ is to be evaluated as $y^* + t\,\delta y$, where $t \in [0, 1]$.

Consider the single-variable function defined by

$$f(t) = G(y^* + t\,\delta y).$$

From Taylor's theorem for a function of a single variable we have

$$f(t) = f(0) + f'(0)t + f''(0)\frac{t^2}{2!} + \cdots + f^r(0)\frac{t^r}{r!} + \frac{1}{(r+1)!} f^{r+1}(\bar{\beta})t^{r+1}$$

where $\bar{\beta} \in (0, t)$ and $f^r(t) \triangleq d^r/dt^r[f(t)]$.

From the definition of $f(\cdot)$ we have

$$f(0)=G(y^*),$$

$$f'(0)=\frac{\partial G(y^*)}{\partial y_1}\delta y_1+\cdots+\frac{\partial G(y^*)}{\partial y_m}\delta y_m$$

$$=U[G(y^*)],$$

$$f''(0)=\frac{\partial^2 G(y^*)}{\partial y_1 y_1}\delta y_1\delta y_1+\frac{\partial^2 G(y^*)}{\partial y_1 y_2}\delta y_1\delta y_2+\cdots+\frac{\partial^2 G(y^*)}{\partial y_m y_m}\delta y_m\delta y_m$$

$$=U^2[G(y^*)],$$

$$\vdots$$

$$f^r(0)=U^r[G(y^*)],$$

$$f^{r+1}(\bar{\beta})=U^{r+1}[G(y^*+\bar{\beta}\,\delta y)].$$

Employing the above relations and evaluating $f(\cdot)$ at $t=1$, we have

$$f(1)=G(y^*+\delta y)=G(y^*)+U[G(y^*)]+\frac{1}{2!}U^2[G(y^*)]+\cdots$$
$$+\frac{1}{r!}U^r[G(y^*)]+\frac{1}{(r+1)!}U^{r+1}[G(y^*+\bar{\beta}\,\delta y)]$$

for some $\bar{\beta}\in(0,1)$.

Taking $\delta y=\alpha\,\hat{\delta y}$, where $\hat{\delta y}$ is a unit vector in the direction of δy and α is a nonnegative real number, and taking $\beta=\bar{\beta}\alpha$, the remainder term becomes

$$R=\frac{1}{(r+1)!}U^{r+1}[G(y^*+\beta\,\hat{\delta y})]$$

for some $\beta\in(0,\alpha)$. ■

In particular, the first- and second-order expansions of Taylor's theorem give us the following corollaries.

Corollary 1.1 Let $G(\cdot): E^m \to E^1$ be C^1 for every point $y^* + \alpha\hat{\delta y} \in B$. Then

$$G(y^* + \alpha\hat{\delta y}) - G(y^*) = \frac{\partial G(y^* + \beta\hat{\delta y})}{\partial y}\delta y \tag{1.13}$$

for some $\beta \in (0, \alpha)$ where $\delta y = \alpha\hat{\delta y}$.

Corollary 1.2 Let $G(\cdot): E^m \to E^1$ be C^2 for every point $y^* + \alpha\hat{\delta y} \in B$. Then

$$G(y^* + \alpha\hat{\delta y}) - G(y^*) = \frac{\partial G(y^*)}{\partial y}\delta y + \tfrac{1}{2}\delta y^T \frac{\partial^2 G(y^* + \beta\hat{\delta y})}{\partial y^2}\delta y \tag{1.14}$$

for some $\beta \in (0, \alpha)$ where $\delta y = \alpha\hat{\delta y}$.

Theorem 1.3 (First-Order Approximation Theorem) Let $G(\cdot): E^m \to E^1$ be C^1 for every point $y^* + \alpha\hat{\delta y} \in B$. Then

$$G(y^* + \alpha\hat{\delta y}) - G(y^*) = \frac{\partial G(y^*)}{\partial y}\delta y + R(\alpha) \tag{1.15}$$

where $\lim_{\alpha \to 0}[R(\alpha)/\alpha] = 0$ and $\delta y = \alpha\hat{\delta y}$.

Proof We obtain (1.15) from Corollary 1.1 by adding and subtracting the term $[\partial G(y^*)/\partial y]\delta y$ to (1.13). By so doing, the remainder term must satisfy

$$R(\alpha) = \left[\frac{\partial G(y^* + \beta\hat{\delta y})}{\partial y} - \frac{\partial G(y^*)}{\partial y}\right]\alpha\hat{\delta y}$$

where $\beta \in (0, \alpha)$. It follows that

$$\frac{R(\alpha)}{\alpha} = \left[\frac{\partial G(y^* + \beta\hat{\delta y})}{\partial y} - \frac{\partial G(y^*)}{\partial y}\right]\hat{\delta y}.$$

But since $G(\cdot)$ is C^1 for every point $y^* + \beta\hat{\delta y}$, $\beta \in (0, \alpha)$, we have that

$$\lim_{\alpha \to 0}\left[\frac{\partial G(y^* + \beta\hat{\delta y})}{\partial y} - \frac{\partial G(y^*)}{\partial y}\right] = 0.$$

Since $\hat{\delta y}$ is a given nonzero vector, it follows that $\lim_{\alpha \to 0}[R(\alpha)/\alpha] = 0$. ■

Theorem 1.4 (Second-Order Approximation Theorem) Let $G(\cdot): E^m \rightarrow E^1$ be C^2 for every point $y^* + \alpha\hat{\delta y} \in B$. Then

$$G(y^* + \alpha\hat{\delta y}) - G(y^*) = \frac{\partial G(y^*)}{\partial y}\delta y + \tfrac{1}{2}\delta y^T \frac{\partial^2 G(y^*)}{\partial y^2}\delta y + R(\alpha^2) \tag{1.16}$$

where $\lim_{\alpha \rightarrow 0}[R(\alpha^2)/\alpha^2] = 0$ and $\delta y = \alpha\hat{\delta y}$.

Proof We obtain (1.16) from Corollary 1.2 by adding and subtracting the term $\frac{1}{2}\delta y^T[\partial^2 G(y^*)/\partial y^2]\delta y$ to (1.14). By so doing, the remainder term must satisfy

$$R(\alpha^2) = \alpha\hat{\delta y}^T\left[\frac{\partial^2 G(y^* + \beta\hat{\delta y})}{\partial y^2} - \frac{\partial^2 G(y^*)}{\partial y^2}\right]\alpha\hat{\delta y}$$

where $\beta \in (0, \alpha)$. Since $G(\cdot)$ is C^2 for every point $y^* + \beta\hat{\delta y}$, $\beta \in (0, \alpha)$, we have that

$$\lim_{\alpha \rightarrow 0}\left[\frac{\partial^2 G(y^* + \beta\hat{\delta y})}{\partial y^2} - \frac{\partial^2 G(y^*)}{\partial y^2}\right] = 0.$$

Since $\hat{\delta y}$ is a given nonzero vector, it follows that $\lim_{\alpha \rightarrow 0}[R(\alpha^2)/\alpha^2] = 0$. ■

The implicit function theorem is concerned with the existence and uniqueness of solutions to systems of equality constraints. Consider a system of n equations in $n+s$ unknowns in the form

$$g_i(x_1, \ldots, x_n, u_1, \ldots, u_s) = 0 \qquad \forall i \in N = \{1, \ldots, n\} \tag{1.17}$$

or, in vector notation,

$$g(x, u) = 0 \tag{1.18}$$

where $g(\cdot): E^n \times E^s \rightarrow E^n$, $x = [x_1, \ldots, x_n]^T$ and $u = [u_1, \ldots, u_s]^T$. We are frequently concerned with the general problem of knowing when (1.18) can be solved for x as a function of u. The following theorem gives sufficient conditions under which this can be done. The theorem is not "constructive," however, in that no mechanism for determining x as a function of u is given, only the conditions under which the determination is possible.

Theorem 1.5 (Implicit Function Theorem) Let (x^*, u^*) be a point in $E^n \times E^s$ with the following properties:

(i) the functions $g_i(\cdot)$ $\forall i \in N$ are continuous with continuous partial derivatives $\partial g_i/\partial x$ in a ball B about (x^*, u^*);

(ii) $g_i(x^*, u^*) = 0$ $\forall i \in n$;

(iii) the Jacobian

$$J = \det\left[\frac{\partial g(x^*, u^*)}{\partial x}\right] = \det\left[\frac{\partial g_i(x^*, u^*)}{\partial x_j}\right] \neq 0$$

where $i, j = 1, \ldots, n$.

Then there exists a vector-valued function $\xi(\cdot): E^s \to E^n$ whose components are single-valued and continuous in a ball B_{u^*} about u^*, with the following properties:

(i) the points $[\xi(u), u]$, $u \in B_{u^*}$, are in B and satisfy the equations $g_i[\xi(u), u] = 0$ $\forall i \in N$;

(ii) there exists a constant ε such that for each u in B_{u^*} the point $[\xi(u^*), u^*]$ is the only solution (x, u) of the equations $g_i(x, u) = 0$ $\forall i \in N$ satisfying the inequalities

$$\|x - \xi(u^*)\| < \varepsilon$$

where $\|\cdot\|$ is the Euclidean norm of $(\cdot)$;

(iii) $x^* = \xi(u^*)$;

(iv) In a sufficiently small ball B_{u^*}, the functions $\xi_i(\cdot)$, $i = 1, \ldots, n$, have continuous partial derivatives of as many orders as are possessed by the functions $g_i(\cdot)$ in the ball B. In other words, if $g_i(\cdot): B \to E^1$ is C^k on B, then $\xi_i(\cdot): B_{u^*} \to E^1$ is C^k on B_{u^*}.

Proof Bliss (1946) or Hestenes (1966). ■

The theorem of the alternative (Mangasarian, 1969) deals with the existence of solutions to systems of linear equalities and inequalities. In the following theorem (and throughout the text) equalities and inequalities such as $>$, $=$, $\geqq$, when applied to vectors, are to be interpreted as applying to all components of the vector.

Theorem 1.6 (Theorem of the Alternative) Let A, B, and D be given matrices and let e, η, μ, and λ be vectors of appropriate dimension. Then either:

(i) $Ae>0$, $Be\geqq 0$, $De=0$ has a solution e;

or

(ii) $\eta^T A+\mu^T B+\lambda^T D=0$ has a solution η, μ, λ with $\mu\geqq 0$, $\eta\geqq 0$, $\eta\neq 0$,

but never both.

Proof (See Mangasarian, 1969). ■

As a consequence of the theorem of the alternative we have Farkas' lemma (Mangasarian, 1969; Farkas, 1902).

Theorem 1.7 (Farkas' Lemma) Let C be the (finite convex) cone generated by the vectors $y^1 \cdots y^q$ in E^m and let C^* be the polar cone to C. Then $y\in C$ if and only if $y^T z\geqq 0\ \forall z\in C^*$.

Proof Let B be the $q\times m$ matrix whose rows are the vectors $y^1 \cdots y^q$:

$$B=\begin{bmatrix} (y^1)^T \\ \vdots \\ (y^q)^T \end{bmatrix}.$$

Then the cone C is given by

$$C=\{y\in E^m \mid y^T=\mu^T B,\ \mu\geqq 0\} \tag{1.19}$$

and the polar cone C^* is given by

$$C^*=\{z\in E^m \mid Bz\geqq 0\}. \tag{1.20}$$

Now consider the system of inequalities

$$y^T z<0 \tag{1.21}$$

$$Bz\geqq 0. \tag{1.22}$$

From the theorem of the alternative (with $y^T=-A$, $D=0$, and $z=e$) either (1.21)–(1.22) has a solution z or there exists a number $\eta>0$ (which may be taken as $\eta=1$ in this case) and a vector $\mu\in E^q$, with $\mu\geqq 0$, such that

$$y^T=\mu^T B, \tag{1.23}$$

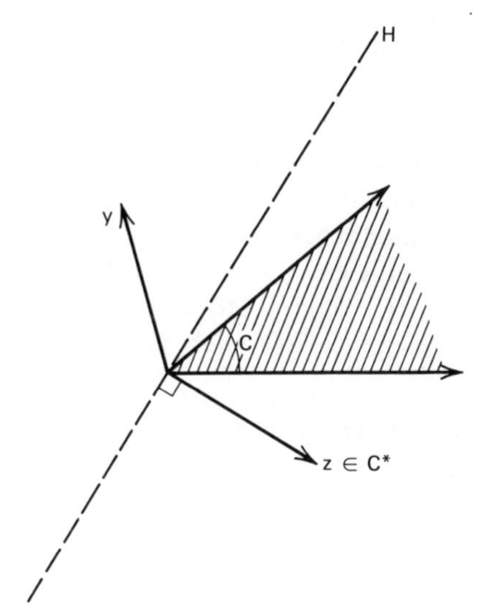

Figure 1.7. Separating hyperplane.

that is, $y \in C$. Thus either there exists a vector $z \in C^*$ such that $y^T z < 0$ or $y \in C$, but not both. ■

Farkas' lemma is also known as the theorem of the separating hyperplane. As illustrated in Figure 1.7, suppose (1.23) has no solution $\mu \geqq 0$, that is, $y \notin C$. Then there exists a vector $z \in C^*$ such that $y^T z < 0$. This means that the hyperplane orthogonal to z, $H = \{y \in E^m | y^T z = 0\}$, has the cone C on one side and the point y on the other.

1.5 EXERCISES

1.1 Determine the global minimum for the function $G(y) = y^3 - 27y + 1$, where $y \in Y = \{y \in E^1 | 10 - y \geqq 0 \text{ and } 10 + y \geqq 0\}$ by examining the plot of $G(\cdot)$ as a function of y.

1.2 Determine the global minimum and maximum for the function $G(y) = y_1^2 - 2y_1 y_2 + 4y_2^2$, where $y \in Y = \{y \in E^2 | -10 \leqq y_i \leqq 10,\ i = 1,2\}$, by direct evaluation.

1.3 Determine the global minimum for the function $G(y) = y_1^2 + y_2^2$, where $y \in Y = \{y \in E^2 | 6y_2 + 4y_1 - 3 \geqq 0\}$, by examining lines of constant $G(\cdot)$ in E^2.

1.4 Determine the global minimum and maximum for the function $G(y) = -(y_1^2+y_2^2)$, where $y \in Y = \{y \in E^2 \mid y_1+3y_2-6=0,\ 5-y_2 \geqq 0,\ 5-y_1 \geqq 0$, and $y_1+2 \geqq 0\}$, by examining lines of constant $G(\cdot)$ in E^2.

1.5 On a certain highway, automobile speed, V, in miles per hour, is restricted to $40 \leqq V \leqq 70$. The fuel consumption rate, F, in gallons per hour, for a certain car has been found experimentally to be $F = 0.03\ V^3/(1000\eta)$ where η is the engine efficiency. For the speed range $40 \leqq V \leqq 70$, experiments yield $\eta = 0.9-(V-55)^2/1000$. Determine the minimum amount of fuel required to travel 300 miles at constant speed.

1.6 For the function $\phi(y) = y_1 e^{2y_2} + y_2 \cos(y_3)$, determine $\partial\phi/\partial y$ and $\partial^2\phi/\partial y^2$.

1.7 Determine the rank of the following matrices:

(a)

$$A = \begin{bmatrix} -1 & 3 & 1 \\ 2 & 1 & 5 \\ 0 & -2 & -2 \end{bmatrix}$$

(b)

$$B = \begin{bmatrix} 2 & 1 & -2 & 0 & 1 \\ 0 & 1 & 2 & 2 & 2 \\ -1 & 0 & 2 & 1 & 1 \end{bmatrix}$$

(c)

$$C = \begin{bmatrix} -1 & 2 & -5 & 4 \\ 1 & 1 & -1 & -1 \\ 2 & 1 & 0 & -3 \end{bmatrix}.$$

1.8 Determine the cone C in E^2 generated by the vectors $y^1 = (1,0)^T$, $y^2 = (1,1)^T$. Determine the polar cone C^* and the dual cone C_D. Draw C, C^*, and C_D on one plot as cones with their vertices at the origin.

Chapter Two

MINIMIZATION OF A SCALAR COST

2.1 PRELIMINARIES

Only for some very special systems is it possible to apply necessary conditions and sufficient conditions for the determination of a *global* minimum of $G(\cdot)$ directly (e.g., linear systems). In general, our analytical methods are confined to the determination of *local* minima.

DEFINITION 2.1 The function $G(\cdot): Y \to E^1$ is said to have a *local minimum* at $y^* \in Y$ if and only if there exists a ball B centered at y^* such that

$$G(y^*) \leqq G(y^* + \delta y) \tag{2.1}$$

for every point $y^* + \delta y \in B \cap Y$. If $G(y^*) < G(y^* + \delta y)$ for all nonzero $\delta y \in E^m$ such that $y^* + \delta y \in B \cap Y$, then $G(\cdot)$ is said to take on a *proper local minimum* at y^*.

It follows from this definition that, if all the local minima for $G(\cdot)$ can be found, then the global minimum of $G(\cdot)$ (if it exists) must be among them.

We employ the definition of a local minimum in the limiting case as the radius of the ball B used in the definition approaches zero. Let $B(\alpha)$ denote a ball centered at y^* with radius α. During the shrinking process we must maintain $y^* + \delta y \in B(\alpha) \cap Y$. This requires a functional dependence of δy on α. We use

$$\delta y = \alpha \, \delta y(\alpha),$$

noting that this form provides not only for the shrinking of δy as $\alpha \to 0$, but also for directional changes in δy that may be required to keep $y^* + \delta y \in B(\alpha) \cap Y$.

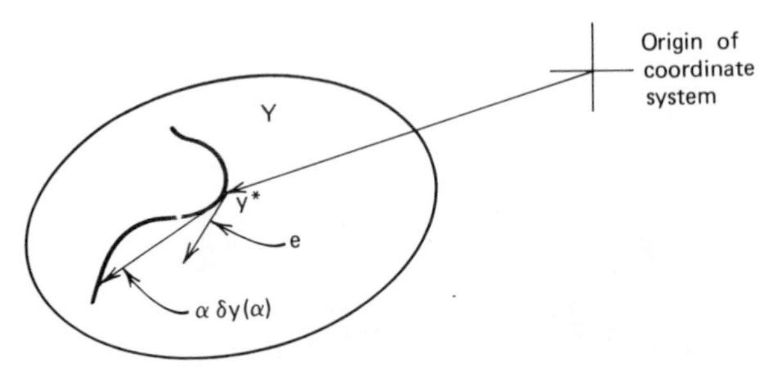

Figure 2.1. A tangent vector to Y.

The definition of a local minimum inherently involves the set Y. This set Y might be a surface in E^m (a set of dimension less than m such as the boundary of a ball), but may also be a set of full dimension (m) in E^m, for example, a region bounded by a surface. The directional relationship of one point in a set to another is expressed in terms of a tangent vector. We need a precise definition of a tangent vector to a set Y, which includes our intuitive notions, but which avoids pathological difficulties that arise if we depend on a heuristic definition only.

The definition of a vector tangent to a set is built from our intuitive notion of a tangent to a curve. Consider the vector formed by connecting two points of any curve lying entirely in Y (see Figure 2.1) and then let one point approach the other. The connecting vector shrinks in magnitude and may rotate during the shrinking process. A tangent vector to the set Y points in the direction of the connecting vector in the limit as the two points become coincident.

DEFINITION 2.2 Let $Y \subseteq E^m$ be locally arcwise connected at $y^* \in \overline{Y}$. We say that a vector $e \in E^m$ is *tangent* to Y at y^* if and only if there exists a positive number γ and a continuous function $\delta y(\cdot):(0,\gamma) \to E^m$ such that:

(i) $y^* + \alpha\,\delta y(\alpha) \in Y$ for all $\alpha \in (0,\gamma)$;

(ii) $\delta y(\alpha) \to e$ as $\alpha \to 0$.

Note that, if e is a tangent vector to Y, then so is βe for all real numbers $\beta \geqq 0$.

> ***Example 2.1*** We note that the definition of a tangent vector is stated with respect to a set Y. In Figure 2.2a a vector e at the boundary of a ball ∂B and normal to the radius vector of the ball is tangent to both the ball ($Y = B$) and to the boundary of the ball ($Y = \partial B$). If the vector e is directed toward the center of the ball (Figure 2.2b), the vector is still tangent to the ball but not to the boundary of the ball.

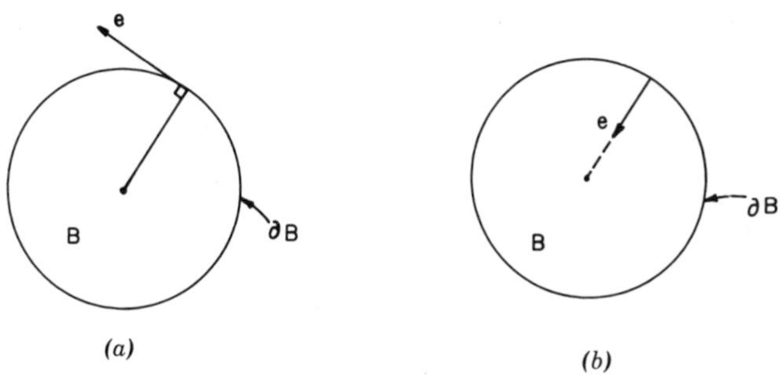

Figure 2.2. Vectors tangent to a ball in E^2.

DEFINITION 2.3 For $Y \subseteq E^m$, with $y^* \in \overline{Y}$, the set of points

$$T = \{e \in E^m | e \text{ tangent to } Y \text{ at } y^*\}$$

is called the *tangent cone* to Y at y^*.

Note: the actual origin for all vectors, cones, gradients, etc., is the zero vector. For illustrative purposes, however, we often consider these as having their origin at the point y^*.

Example 2.2 Let Y be the set

$$Y = \{y \in E^2 | h_1(y) \geqq 0 \text{ and } h_2(y) \geqq 0\}$$

where

$$h_1(y) = y_1 + 1 - y_2$$

$$h_2(y) = y_1 + y_2 + 1.$$

The set Y is shown in Figure 2.3. At point $A = (-1, 0)$ the tangent cone is a convex cone with boundaries $h_1(y) = 0$ and $h_2(y) = 0$. At point B the tangent cone is a closed half-space bounded by $h_1(y) = 0$. At point C interior to Y, the tangent cone is all of E^2.

We now present the basic first-order necessary conditions and sufficient conditions for a local minimum.

Lemma 2.1 (First-Order) Let $y^* \in Y$ be a local minimum point for the function $G(\cdot): Y \to E^1$, which is C^1 on Y. It is necessary that

$$\frac{\partial G(y^*)}{\partial y} e \geqq 0 \tag{2.2}$$

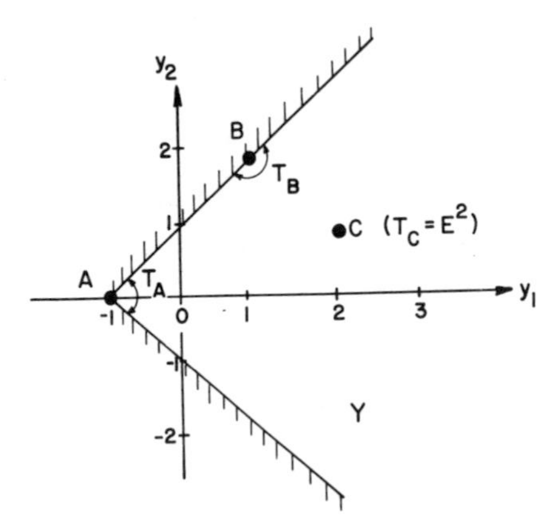

Figure 2.3. Tangent cones for Example 2.2.

for all $e \in T$, the tangent cone to Y at y^*. Furthermore, the condition

$$\frac{\partial G(y^*)}{\partial y} e > 0 \tag{2.3}$$

for all nonzero $e \in T$ is sufficient for a proper local minimum at y^*.

Proof If $y^* \in Y$ is a local minimum for $G(\cdot)$, then from Definition 2.1

$$G[y^* + \alpha\,\delta y(\alpha)] - G(y^*) \geqq 0 \tag{2.4}$$

for all $y^* + \alpha\,\delta y(\alpha) \in B(\alpha) \cap Y$. We now take the limit as $\alpha \to 0$, noting that $\lim_{\alpha\to 0} \alpha\,\delta y(\alpha) = (\lim_{\alpha \to 0} \alpha) \lim_{\alpha \to 0} \delta y(\alpha)$ since both limits exist. From the definitions of a tangent vector and a minimum, it follows that at a minimal point

$$\lim_{\alpha \to 0} [G(y^* + \alpha e) - G(y^*)] \geqq 0$$

for all $e \in T$. This may be used as an alternate definition of a local minimum. From the first-order approximation theorem,

$$\lim_{\alpha \to 0} \left[\frac{\partial G(y^*)}{\partial y} \alpha e + R(\alpha) \right] \geqq 0$$

for all $e \in T$. Dividing by $\alpha > 0$, we have in the limit that

$$\frac{\partial G(y^*)}{\partial y} e \geqq 0$$

for all $e \in T$.

To establish the sufficiency of (2.3) suppose

$$\frac{\partial G(y^*)}{\partial y} e > 0$$

for all nonzero $e \in T$. From Definition 2.2, for each tangent vector e there exists a function $\delta y(\cdot)$ and a positive number γ such that

$$y^* + \alpha \delta y(\alpha) \in Y$$

for all $\alpha \in (0, \gamma)$ where $\delta y(\alpha) \to e$ as $\alpha \to 0$. From the first-order approximation theorem we have

$$G[y^* + \alpha \delta y(\alpha)] - G(y^*) = \alpha \left[\frac{\partial G(y^*)}{\partial y} \delta y(\alpha) + \frac{R(\alpha)}{\alpha} \right].$$

Note that, if $[\partial G(y^*)/\partial y]\delta y(\alpha) = 0$, then the sign of the difference is determined by the sign of $R(\alpha)$. Otherwise, for sufficiently small $\alpha > 0$, $\delta y(\alpha) \to e \in T$, $R(\alpha)/\alpha \to 0$, and the sign of the difference is determined by the sign of $[\partial G(y^*)/\partial y]e$, provided that this term is not zero. Thus (2.3) implies that y^* is a proper local minimum point for $G(\cdot)$. ■

Note that, if $T = E^m$, it follows that e may be selected arbitrarily and an immediate consequence of Lemma 2.1 is the familiar result that $\partial G(y^*)/\partial y = 0$ for an unconstrained local minimum (or maximum). The following examples illustrate some implications of the lemma when $T \subset E^m$.

Example 2.3 Consider the problem of maximizing the function $H(\cdot) = y_1^2 + y_2^2$ with Y defined by $h_1(\cdot) = 4 - (y_1 - 1)^2 - y_2^2 \geqq 0$. In order to maximize $H(\cdot)$ we minimize $G(\cdot) = -y_1^2 - y_2^2$. The necessary condition (2.2) is satisfied at points $(-1, 0)$ and $(3, 0)$ as illustrated in Figure 2.4. The tangent cone at $(-1, 0)$ is the half-space to the right of and including the line $y_1 = -1$ and the tangent cone at $(3, 0)$ is the half-space to the left of and including the line $y_1 = 3$.

Note that neither $(-1, 0)$ nor $(3, 0)$ satisfies the sufficient condition (2.3) since there exists an $e \in T$ that is perpendicular to the gradient of G. The point $(-1, 0)$ is not a local minimal point, but $(3, 0)$ is a local minimal point as can be verified from the definition of a local minimum. Lines of constant cost (two of which are illustrated in Figure 2.4) aid in understanding the significance of the latter observations.

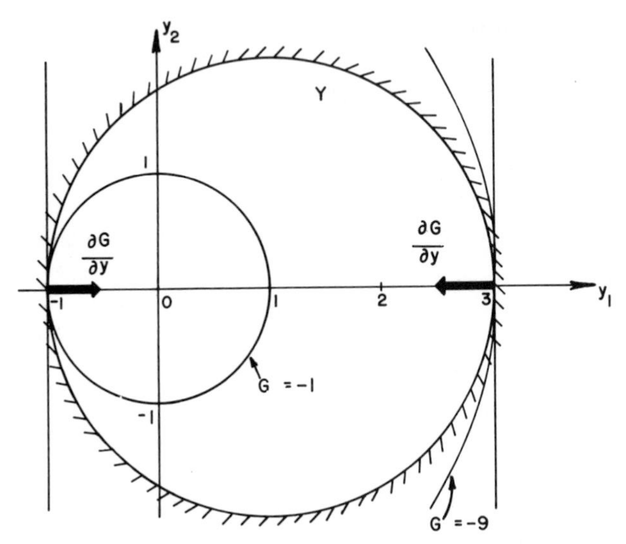

Figure 2.4. Geometry for Example 2.3.

The restriction "$e \in T$" in Lemma 2.1 is required for the lemma to hold. In particular,

$$\frac{\partial G(y^*)}{\partial y}\delta y > 0$$

for all nonzero $\delta y \in B \cap Y$, where B is a ball about y^*, is *not* sufficient to guarantee that y^* is a minimal point. At $(-1,0)$ in Example 2.3, the above condition is satisfied, but $(-1,0)$ is not a minimal point.

Example 2.4 Consider the problem of maximizing the function $H(\cdot) = y_1^2 + y_2^2$ with Y defined by $h_1(\cdot) = y_1 + 1 - y_2 \geqq 0$, $h_2(\cdot) = y_2 + y_1 + 1 \geqq 0$, and $h_3(\cdot) = 2 - y_1 \geqq 0$. Again we minimize $G(\cdot) = -y_1^2 - y_2^2$. The necessary condition (2.2) is satisfied at points $(-1,0)$, $(-\frac{1}{2}, \frac{1}{2})$, $(-\frac{1}{2}, -\frac{1}{2})$, $(2,0)$, $(2,3)$, and $(2,-3)$, as shown in Figure 2.5.

The sufficient condition (2.3) is satisfied at points $(-1,0)$, $(2,3)$, and $(2,-3)$, which are also the only local minimal points. The function $G(\cdot)$ takes on a (global) minimum at $(2,3)$ and $(2,-3)$. Note that, if $h_3(y) \geqq 0$ were not specified as a constraint, then Y would be unbounded and $G(\cdot)$ would have no global minimum.

The possibility of equality in (2.2) leads us to a further definition.

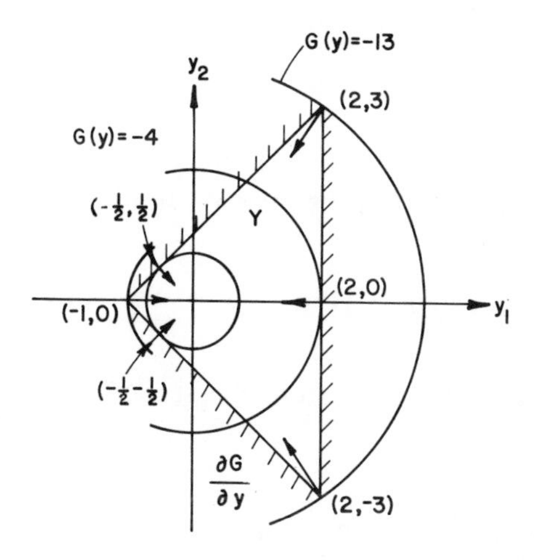

Figure 2.5. Geometry for Example 2.4.

DEFINITION 2.4 The function $G(\cdot): Y \to E^1$ is said to be *stationary* at a point $y^* \in Y$ if and only if

$$\frac{\partial G(y^*)}{\partial y} e = 0 \tag{2.5}$$

for all $e \in T$. The point y^* is said to be a *stationary point.*

We may now employ the second-order approximation theorem to examine stationary minimal points.

Lemma 2.2 (Second-Order) Let $y^* \in Y$ be a stationary local minimum point for a function $G(\cdot): Y \to E^1$, which is C^2 on Y. It is necessary that

$$\tfrac{1}{2} e^T \frac{\partial^2 G(y^*)}{\partial y^2} e \geqq 0 \tag{2.6}$$

for all $e \in T$. Furthermore, the condition

$$\tfrac{1}{2} e^T \frac{\partial^2 G(y^*)}{\partial y^2} e > 0 \tag{2.7}$$

for all nonzero $e \in T$ is sufficient for a proper stationary local minimum.

Proof If $y^* \in Y$ is a local minimum point for $G(\cdot)$, then from Definition 2.1 there exists a positive number γ, a continuous function

$\delta y(\cdot):(0,\gamma)\rightarrow E^m$, and a sequence of balls $B(\alpha)$ centered at y^* such that

$$G[y^*+\alpha\,\delta y(\alpha)]-G(y^*)\geqq 0$$

for all $y^*+\alpha\,\delta y(\alpha)\in B(\alpha)\cap Y$ and for all $\alpha\in(0,\gamma)$. In view of the definitions of the tangent vector and tangent cone, we have in the limit as $\alpha\rightarrow 0$ that

$$\lim_{\alpha\rightarrow 0}\left[G(y^*+\alpha e)-G(y^*)\right]\geqq 0 \tag{2.8}$$

for all $e\in T$. Since y^* is stationary, we have from the second-order approximation theorem

$$\lim_{\alpha\rightarrow 0}\left[\tfrac{1}{2}\alpha e^T\frac{\partial^2 G(y^*)}{\partial y^2}\alpha e+R(\alpha^2)\right]\geqq 0$$

for all $e\in T$. Dividing by α^2 and taking the limit as $\alpha\rightarrow 0$, we obtain (2.6).

To establish the sufficiency of (2.7), suppose that

$$\tfrac{1}{2}e^T\frac{\partial^2 G(y^*)}{\partial y^2}e>0$$

for all nonzero $e\in T$. The condition $e\in T$ implies

$$y^*+\alpha\,\delta y(\alpha)\in Y$$

for all $\alpha\in(0,\gamma)$ where $\delta y(\alpha)\rightarrow e$ as $\alpha\rightarrow 0$. From the second-order approximation theorem and the definition of a stationary point,

$$G[y^*+\alpha\,\delta y(\alpha)]-G(y^*)=\alpha^2\left[\tfrac{1}{2}\delta y(\alpha)^T\frac{\partial^2 G(y^*)}{\partial y^2}\delta y(\alpha)+\frac{R(\alpha^2)}{\alpha^2}\right]$$

where $R(\alpha^2)/\alpha^2\rightarrow 0$ as $\alpha\rightarrow 0$. Since $\delta y(\alpha)\rightarrow e$ as $\alpha\rightarrow 0$, we obtain

$$\lim_{\alpha\rightarrow 0}\left\{\frac{G[y^*+\alpha e]-G[y^*]}{\alpha^2}\right\}=\tfrac{1}{2}e^T\frac{\partial^2 G(y^*)}{\partial y^2}e>0,$$

which implies that y^* is a proper stationary local minimal point for $G(\cdot)$.

■

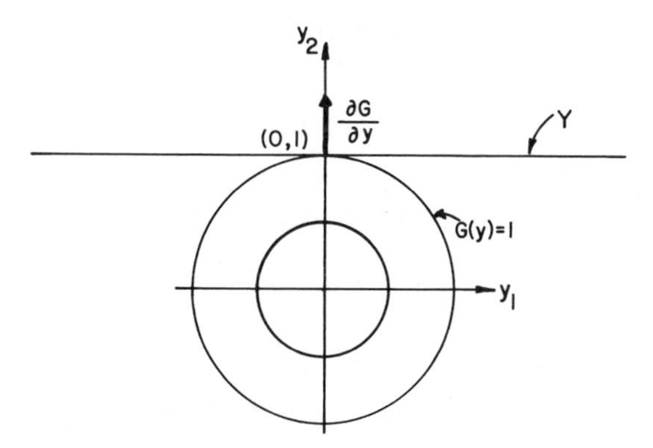

Figure 2.6. Geometry for Example 2.5.

Example 2.5 Let $G(\cdot)=y_1^2+y_2^2$ with a single equality constraint $g(\cdot)=y_2-1=0$. The domain of $G(\cdot)$, some constant cost curves, and the gradient of $G(\cdot)$ at the point $(0,1)$ are illustrated in Figure 2.6. In this case the point $(0,1)$ is the only point in Y that satisfies the necessary condition (2.2). It is also a stationary point (note $T=Y$ for any $y\in Y$). We obtain

$$\left.\frac{\partial^2 G}{\partial y^2}\right|_{(0,1)}=\begin{bmatrix}2 & 0\\ 0 & 2\end{bmatrix}.$$

Any nonzero tangent vector in the tangent cone at $(0,1)$ may be represented by $(a,0)^T$, where a is a nonzero real number. Thus

$$\tfrac{1}{2}e^T\left.\frac{\partial^2 G}{\partial y^2}\right|_{(0,1)}e=a^2>0.$$

Hence, the point $(0,1)$ satisfies the sufficient condition (2.7) and is a proper stationary local minimal point.

Example 2.6 Consider the problem of minimizing $G(\cdot)=y^2$ where $Y=E^1$. For any point of Y the first-order necessary condition (2.2) cannot be satisfied unless

$$\frac{\partial G(y^*)}{\partial y}=0.$$

Obviously, this latter condition is satisfied at the point $y=0$, which is also stationary. The second-order sufficiency condition (2.7) is also satisfied at $y=0$ so that the point is a proper stationary local minimum.

The possibility of equality in (2.6) leads us to one further definition.

DEFINITION 2.5 The function $G(\cdot): Y \to E^1$ is said to be *singular* at $y^* \in Y$ if and only if y^* is stationary and if, in addition,

$$e^T \frac{\partial^2 G(y^*)}{\partial y^2} e = 0 \tag{2.9}$$

for all $e \in T$. The point y^* is called a *singular point*.

Example *2.7* Let $G(\cdot)=y^3$ and $Y=E^1$. In this case, the first- and second-order necessary conditions are satisfied at the point $y=0$. This point is also singular. Note that in this case $y=0$ is not a local minimal point.

2.2 REGULAR POINTS AND NORMAL POINTS

Let Y be defined by $g(y)=0$ and $h(y) \geqq 0$ in a ball B about a point $y^* \in Y$ where $g(\cdot): E^m \to E^n$ and $h(\cdot): E^m \to E^q$ are C^1. Consider points in B given by

$$B \cap Y = \{ y \in B \mid g(y)=0 \text{ and } h(y) \geqq 0 \}.$$

An inequality constraint is said to be *active* at y^* if $h_j(y^*)=0$ and *inactive* if $h_j(y^*)>0$. For convenience reorder the indices, if necessary, so that $Q^* \triangleq \{ j \mid h_j(y^*)=0 \} = \{1,\ldots,q^*\}$ and let $\hat{h}(\cdot)=[h_1(\cdot),\ldots,h_{q^*}(\cdot)]^T$ denote the active inequality constraints at y^*. Let $\delta y(\cdot):(0,\gamma) \to E^m$ generate a vector e tangent to Y at y^*.

From the first-order approximation theorem

$$g[y^* + \alpha \delta y(\alpha)] = g(y^*) + \alpha \frac{\partial g(y^*)}{\partial y} \delta y(\alpha) + R(\alpha)$$

$$\hat{h}[y^* + \alpha \delta y(\alpha)] = \hat{h}(y^*) + \alpha \frac{\partial \hat{h}(y^*)}{\partial y} \delta y(\alpha) + \overline{R}(\alpha)$$

where $y^* + \alpha \delta y(\alpha) \in Y$, $\delta y(\alpha) \to e$ as $\alpha \to 0$, $R(\alpha)/\alpha \to 0$, and $\overline{R}(\alpha)/\alpha \to 0$ as

$\alpha \to 0$. Since $y^* \in Y$ and $y^* + \alpha\delta y(\alpha) \in Y$, we have

$$0 = \alpha \frac{\partial g(y^*)}{\partial y} \delta y(\alpha) + R(\alpha)$$

$$0 \leqq \alpha \frac{\partial \hat{h}(y^*)}{\partial y} \delta y(\alpha) + \bar{R}(\alpha).$$

Dividing by $\alpha > 0$ and taking the limit as $\alpha \to 0$ yields

$$\frac{\partial g(y^*)}{\partial y} e = 0 \tag{2.10}$$

$$\frac{\partial \hat{h}(y^*)}{\partial y} e \geqq 0. \tag{2.11}$$

Thus every tangent vector at y^* must satisfy (2.10) and (2.11). This does not imply, however, that an arbitrary vector e that satisfies (2.10) and (2.11) is tangent to Y at y^*. Consider for example the case where $\partial g_i / \partial y = \partial h_j / \partial y = 0$ $\forall i \in N$, $\forall j \in Q^*$. In such a situation every vector $e \in E^m$ satisfies (2.10) and (2.11).

To illustrate one way that this situation could occur, consider the case in E^2 where the constraint set Y is defined by a single inequality constraint $h_1(\cdot) = y_1^3 - y_2^2 \geqq 0$. At the origin $h_1(0,0) = 0$, so that the constraint is active, but $\partial h_1 / \partial y = [0, 0]$. Thus every vector $e \in E^2$ satisfies (2.11). But the constraint boundary $h_1(y) = 0$ has a cusp at the origin and only vectors of the form $e^T = [k, 0]$, $k \geqq 0$, are tangent to Y at the origin. If we impose, explicitly, the additional constraint $h_2(\cdot) = y_1 \geqq 0$ in the definition of Y, then e is a tangent vector if and only if e satisfies (2.11).

DEFINITION 2.6 A point $y^* \in Y$ is a *regular point* of Y if and only if

(i) the set $B \cap Y$ contains neighboring points to y^*;
(ii) $g(\cdot)$ and $h(\cdot)$ are C^1 on $B \cap Y$;
(iii) the tangent cone T to Y at y^* consists of exactly those vectors e satisfying (2.10) and (2.11). If N and Q^* are empty, then $T = E^m$.

The set Y is said to be a regular set if every point of Y is regular. If N and Q^* are both empty at y^*, then y^* is a regular point.

DEFINITION 2.7 A point $y^* \in Y$ is a *normal point* of Y if and only if

(i) the set $B \cap Y$ contains neighboring points to y^*;
(ii) $g(\cdot)$ and $h(\cdot)$ are C^1 on $B \cap Y$;

(iii) the vectors

$$\frac{\partial g_i(y^*)}{\partial y} \qquad \forall i \in N$$

$$\frac{\partial h_j(y^*)}{\partial y} \qquad \forall j \in Q^*$$

are linearly independent.

The set Y is said to be a normal set if every point of Y is normal. If N and Q^* are both empty at y^*, then y^* is a normal point. Points that are not normal are called abnormal.

Lemma 2.3 If y^* is a normal point of Y, then y^* is a regular point of Y.

Proof ***(Hestenes, 1966)*** If N and Q^* are both empty, then the lemma follows from the definitions. If N and Q^* are not both empty, then for notational purposes we define

$$f_i(y) = \begin{cases} g_i(y) & i \in N = \{1, \dots, n\} \\ h_j(y) & j \in Q^* = \{1, \dots, q^*\},\ i = n+j, \dots, p = n+q^*. \end{cases}$$

From the definition of normality the $p \times m$ matrix

$$\frac{\partial f}{\partial y} = \begin{bmatrix} \dfrac{\partial f_1(y^*)}{\partial y} \\ \vdots \\ \dfrac{\partial f_p(y^*)}{\partial y} \end{bmatrix}$$

has rank p. Thus the linear transformation represented by $\partial f / \partial y$ maps E^m *onto* E^p. Let $\{x^1, \dots, x^p\}$ be a basis for E^p generated by applying the linear transformation to certain vectors $e^i \in E^m$, $i = 1, \dots, p$. Then the vectors $x^i \in E^p$, $i = 1, \dots, p$ given by

$$[x^1, \dots, x^p] = \frac{\partial f}{\partial y}[e^1, \dots, e^p]$$

are linearly independent. Therefore the $p \times p$ determinant

$$\left| \frac{\partial f}{\partial y}[e^1, \dots, e^p] \right| \neq 0. \tag{2.12}$$

This result is used in conjunction with the implicit function theorem to show that, if y^* is a normal point of Y, then any vector e that satisfies (2.10) and (2.11), that is,

$$\frac{\partial f_i(y^*)}{\partial y}e=0 \qquad i=1,\dots,n$$

$$\frac{\partial f_i(y^*)}{\partial y}e\geqq 0 \qquad i=n+1,\dots,p, \tag{2.13}$$

is in fact tangent to Y at y^*. Specifically, we show that there exists a curve

$$y(\alpha)=y^*+\alpha e+\sum_{k=1}^{p}\beta_k(\alpha)e^k$$

in Y with $y(0)=y^*$ and $y'(0)=e$, where the prime denotes differentiation with respect to α.

Consider the equation

$$F_i(\alpha,\beta_1,\dots,\beta_p)\triangleq f_i\left(y^*+\alpha e+\sum_{k=1}^{p}\beta_k e^k\right)-\alpha\frac{\partial f_i(y^*)}{\partial y}e=0, \tag{2.14}$$

where $i=1,\dots,p$. Since $y^*\in Y$, (2.14) has a solution $\alpha=\beta_k=0$, $k=1,\dots,p$. From the implicit function theorem, with the Jacobian $|\partial F_i/\partial\beta_k|$, $i,k=1,\dots,p$, given by (2.12), there exist functions $\beta_k(\alpha)$ with $\beta_k(0)=0$ that satisfy (2.14) on some interval $-\delta\leqq\alpha\leqq\delta$. Therefore, from (2.13) and (2.14)

$$y^*+\alpha e+\sum_{k=1}^{p}\beta_k(\alpha)e^k\in Y \qquad \text{for } \alpha\in[0,\delta].$$

To complete the proof that e is a tangent vector (i.e., $\alpha e+\sum_{k=1}^{p}\beta_k(\alpha)e^k \to e$ as $\alpha\to 0$), we show that $y'(0)=e$. Differentiating (2.14) with respect to α and evaluating at $\alpha=0$ yields the system of equations

$$\frac{\partial f}{\partial y}[e^1,\dots,e^p]\begin{bmatrix}\beta_1'(0)\\ \vdots\\ \beta_p'(0)\end{bmatrix}=\begin{bmatrix}0\\ \vdots\\ 0\end{bmatrix}.$$

From (2.12) these equations have only the trivial solution $\beta_k'(0)=0$, $k=1,\dots,p$. Thus $y'(0)=e$, which implies that e is tangent to Y at y^*. ■

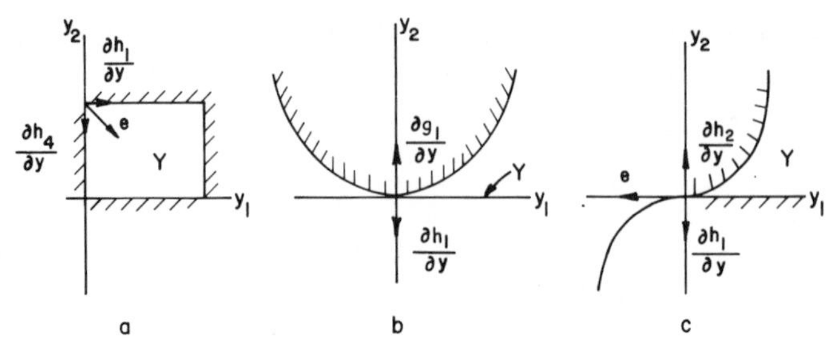

Figure 2.7. Surfaces with regular, irregular, normal, and abnormal points. (*a*) Regular and normal (Example 2.8). (*b*) Regular and abnormal (Example 2.9). (*c*) Irregular and abnormal (Example 2.10).

Example 2.8 Every point $y=(y_1, y_2)\in Y$ with Y defined by

$$h_1(\cdot)=y_1\geqq 0$$

$$h_2(\cdot)=y_2\geqq 0$$

$$h_3(\cdot)=(1-y_1)\geqq 0$$

$$h_4(\cdot)=(1-y_2)\geqq 0,$$

as illustrated in Figure 2.7*a*, is regular and normal. At point (0, 1), for example, every vector e that satisfies (2.11) is in T and the gradients $\partial h_1/\partial y$ and $\partial h_4/\partial y$ are linearly independent.

Example 2.9 The point $(0,0)\in Y$, with Y defined by

$$h_1(\cdot)=y_1^2-y_2\geqq 0$$

$$g_1(\cdot)=y_2=0,$$

as illustrated in Figure 2.7*b*, is regular and abnormal. At (0,0) every vector e that satisfies (2.10) and (2.11) is in T but the gradients $\partial h_1/\partial y$ and $\partial g_1/\partial y$ are linearly dependent.

Example 2.10 The point $(0,0)\in Y$, with Y defined by

$$h_1(\cdot)=y_1^3-y_2\geqq 0$$

$$h_2(\cdot)=y_2\geqq 0,$$

as illustrated in Figure 2.7*c*, is irregular and abnormal. At (0,0) the vector $e \notin T$ satisfies (2.11) and the gradients $\partial h_1/\partial y$ and $\partial h_2/\partial y$ are linearly dependent.

It should be noted that an assumption of regularity for Y is equivalent to an assumption that Y satisfies the Karush-Kuhn-Tucker constraint qualification condition. This condition is frequently cited in the nonlinear programming literature. Until recently, the condition was attributed to Kuhn and Tucker as promulgated in their pioneering work on nonlinear programming (Kuhn and Tucker, 1951). However, as both Kuhn (1976) and Takayama (1974) point out, the condition was developed independently, prior to the Kuhn-Tucker paper, by William Karush as a Master's Thesis (Karush, 1939) at the University of Chicago. For the interested reader, Karush's thesis was never published, but Kuhn (1976) gives an account of Karush's results in crediting him with the initial development of the constraint qualification condition.

It follows from Definition 2.6 that, if the constraint set

$$Y = \{y \in E^m | g(y) = 0 \text{ and } h(y) \geqq 0\}$$

is regular (in a ball about a point $y^* \in Y$), then the tangent cone T to Y at y^* is given by

$$T = \left\{ e \in E^m | \frac{\partial g(y^*)}{\partial y} e = 0 \text{ and } \frac{\partial \hat{h}(y^*)}{\partial y} e \geqq 0 \right\}$$

where $\hat{h}(y^*)$ are the active inequality constraints at y^*.

Consider the case where the system is only subject to the inequality constraints $h(y) \geqq 0$. The set Y is then the *intersection* of the sets $\{y \in E^m | h_j(y) \geqq 0\}$, $j \in Q^*$. With two inequality constraints in E^2, the tangent cone to the constraint set Y is convex, as illustrated in Figure 2.8*a*. Note

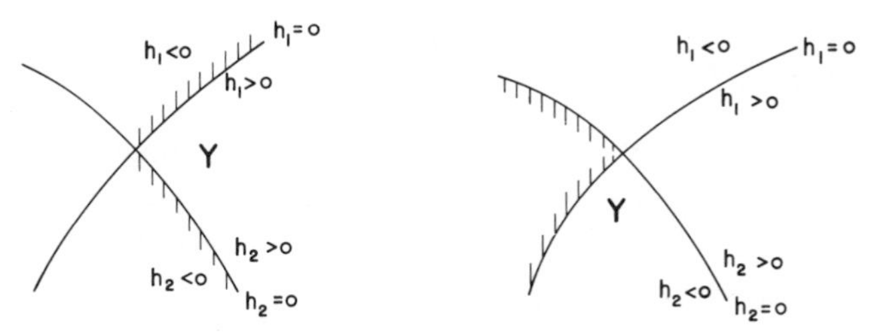

Figure 2.8. A two-dimensional example of inequality contraint sets. (a) $Y = \bigcap_j \{y | h_j(y) \geqq 0\}$. (b) $Y = \bigcup_j \{y | h_j(y) \geqq 0\}$.

that a nonconvex tangent cone would result if Y were taken as the *union* of the sets $\{y \in E^m | h_j(y) \geqq 0\}, j \in Q^*$, as illustrated in Figure 2.8*b*.

The convexity result illustrated in Figure 2.8 is shown to be true in general with the following lemma.

Lemma 2.4 If y^* is a regular point of Y, then T, the set of vectors tangent to Y at y^*, is a convex cone.

Proof If y^* is a regular point of Y, then T is exactly the set of all e such that

$$\frac{\partial g(y^*)}{\partial y} e = 0$$

$$\frac{\partial \hat{h}(y^*)}{\partial y} e \geqq 0.$$

If $e \in T$, these equations imply $\beta e \in T$ for all real numbers $\beta \geqq 0$; thus T is a cone by definition. To demonstrate convexity choose arbitrary $e^1 \in T$, $e^2 \in T$, and let

$$e = (1-\alpha)e^1 + \alpha e^2,$$

where $\alpha \in [0,1]$. Then

$$\frac{\partial g(y^*)}{\partial y} e = (1-\alpha)\frac{\partial g(y^*)}{\partial y} e^1 + \alpha \frac{\partial g(y^*)}{\partial y} e^2 = 0$$

and

$$\frac{\partial \hat{h}(y^*)}{\partial y} e = (1-\alpha)\frac{\partial \hat{h}(y^*)}{\partial y} e^1 + \alpha \frac{\partial \hat{h}(y^*)}{\partial y} e^2 \geqq 0.$$

Therefore, $e \in T$ for all $\alpha \in [0,1]$. ■

The tangent cone T has a simple geometric relationship to the cone generated by the gradient vectors $\partial g/\partial y$, $-\partial g/\partial y$, and $\partial h/\partial y$ associated with the constraint equations. At any point $y^* \in Y$ with $g(\cdot)$ and $h(\cdot) C^1$ at y^* we define the following *constraint cone*:

$$K = \left\{ y \in E^m | y^T = \lambda^T \frac{\partial g(y^*)}{\partial y} + \mu^T \frac{\partial h(y^*)}{\partial y}, \lambda \in E^n, \mu \in E^q, \mu \geqq 0, \mu^T h(y^*) = 0 \right\} \tag{2.15}$$

where

$$\lambda^T \frac{\partial g(y^*)}{\partial y} \triangleq (\beta^T - \gamma^T) \frac{\partial g(y^*)}{\partial y}$$

with $\beta \geqq 0$, $\gamma \geqq 0$. The polar cone to K is given by

$$K^* = \{ e \in E^m \mid y^T e \geqq 0 \ \forall y \in K \}. \tag{2.16}$$

Lemma 2.5 If y^* is a regular point of Y, then the tangent cone T to Y at y^* is given by

$$T = K^*.$$

Proof From the definition of a regular point, $e \in T$ if and only if e satisfies (2.10) and (2.11). From the definition of K^*, $e \in K^*$ if and only if

$$\lambda^T \frac{\partial g(y^*)}{\partial y} e + \mu^T \frac{\partial h(y^*)}{\partial y} e \geqq 0 \tag{2.17}$$

$\forall \lambda \in E^n$ and $\forall \mu \in E^q$ such that $\mu \geqq 0$ and

$$\mu^T h(y^*) = 0. \tag{2.18}$$

For $e \in K^*$ suppose $[\partial g(y^*)/\partial y] e \neq 0$. Then, since λ is arbitrary, it could be chosen so that (2.17) is violated, which would contradict $e \in K^*$. Therefore, if $e \in K^*$, then e satisfies (2.10). Similarly, for $e \in K^*$ suppose $[\partial h_j(y^*)/\partial y] e < 0$ for some $j \in Q^*$. Condition (2.18) and $\mu \geqq 0$ imply $\mu_j = 0 \ \forall j \notin Q^*$ and $\mu_j \geqq 0 \ \forall j \in Q^*$. Therefore, μ could be chosen so that (2.17) is violated, which would contradict $e \in K^*$. Thus if $e \in K^*$, then e must satisfy (2.11). Therefore, $K^* \subseteq T$. On the other hand, if $e \in T$, then e satisfies (2.10)–(2.11), which implies that e satisfies (2.17) $\forall \lambda \in E^n$ and $\forall \mu \geqq 0$ with μ constrained by (2.18). Therefore, $T \subseteq K^*$ and we conclude that $T = K^*$.

2.3 INTERNAL POINTS

Suppose that $y^* \in Y$ is a local minimum point for the function $G(\cdot)$. Then by Lemma 2.1 it is necessary that

$$\frac{\partial G(y^*)}{\partial y} e \geqq 0 \qquad \forall e \in T. \tag{2.19}$$

Note that, should $\partial G(y^*)/\partial y=0$, this condition is satisfied and y^* is a candidate minimal point. In what follows it is generally assumed, in the presence of constraints, that $\partial G(y^*)/\partial y \neq 0$ (i.e., at least one component of the vector is nonzero), although this is by no means a requirement.

We now assume that Y is a regular set so that the tangent vectors $e \in T$ are defined by means of (2.10) and (2.11). We first consider separately the case where Q^* is empty at a given $y^* \in Y$.

DEFINITION 2.8 We say that $y^* \in Y$ is an *internal point* of Y if and only if Q^* is empty.†

At an internal point all inequality constraints are satisfied as strict inequalities (i.e., all inequality constraints are "inactive" at y^*). Note that, if there are no inequality constraints specified, then every point of Y is an internal point. (Q is empty $\Rightarrow Q^*$ is empty.)

2.4 FIRST-ORDER NECESSARY CONDITIONS FOR INTERNAL POINTS

For internal points, because of our assumption that Y is regular, if $e \in T$, then $-e \in T$ and the necessary condition (2.12) is violated unless

$$\frac{\partial G(y^*)}{\partial y} e=0 \qquad \forall e \in T. \tag{2.20}$$

We conclude that, for regular internal points of Y, local minimal points of G are stationary.

At such a point first-order changes in $G(y^*)$ are zero and we are led to the *principle of flat laxity* (Isaacs, 1967). At such a minimum the precise value of the argument of the function is generally not critical. Isaacs notes, "It is highly likely that the flat laxity principle accounts for the astonishing dearth of practical applications of the many elegant mathematical maxima that have appeared throughout scientific history. The right circular cylinder of a given volume has minimal surface area when the height is equal to the diameter. However, tin cans of the proportion are seldom seen on the market shelves."

†We resist the temptation to call these points relative interior points (under the usual mathematical definitions). This can be understood by a simple example. Suppose that N is empty (no equality constraints), $m=2$, and $h_1(y)=y_1-a \geqq 0$, $h_2(y)=a-y_2 \geqq 0$. Then $Y=\{y \in E^2 \colon y_1=a\}$ and the line $y_1=a$ contains no internal points by our definition; however, every point of the line is a relative interior point of the one-dimensional set Y under the usual definitions.

Lemma 2.6 (Regular Internal Points) Let y^* be a regular internal point of Y. Let $G(\cdot): Y \to E^1$ and $g(\cdot): E^m \to E^n$ be C^1 in a neighborhood $B \cap Y$ of y^*. If $G(\cdot)$ takes on a local minimum at y^*, then

$$\operatorname{rank}\left[\frac{\partial g}{\partial y}\right]_{y^*} = \operatorname{rank}\left[\begin{array}{c} \dfrac{\partial G}{\partial y} \\ \dfrac{\partial g}{\partial y} \end{array}\right]_{y^*}. \tag{2.21}$$

Proof Consider the tangent cone T at y^*. Now let T_G be the set of vectors $e \in E^m$ at y^* satisfying (2.20) with the understanding that $T_G = E^m$ if $\partial G(y^*)/\partial y$ is the zero vector. Let $T' = T \cap T_G \subseteq T$. Since (2.20) and (2.10) are homogeneous relations, it follows from linear algebra (Nering, 1963) that T and T' are vector spaces with dimensions

$$\dim(T) = m - \operatorname{rank}\left[\frac{\partial g}{\partial y}\right]_{y^*}$$

$$\dim(T') = m - \operatorname{rank}\left[\begin{array}{c} \dfrac{\partial G}{\partial y} \\ \dfrac{\partial g}{\partial y} \end{array}\right]_{y^*}$$

where $\partial g/\partial y$ is the $n \times m$ matrix with elements $\partial g_i/\partial y_j$, $i = 1, \ldots, n$ (row), $j = 1, \ldots, m$ (column). Thus either $\dim(T) = \dim(T')$ or $\dim(T) = \dim(T') + 1$. Suppose the latter condition holds. This implies that $T' \subset T$ or, more specifically, that there exists a vector e in the tangent cone T at $y^* \in Y$ that does not satisfy (2.20). This contradicts the assumption that y^* is a minimal point; hence, $\dim(T) = \dim(T')$. ■

Thinking in terms of the row vectors of (2.21), we obtain an alternate form of the regular internal point lemma.

Lemma 2.7 Let y^* be a regular internal point of Y. Let $G(\cdot): Y \to E^1$ and $g(\cdot): E^m \to E^n$ be C^1 in a neighborhood $B \cap Y$ of y^*. If $G(\cdot)$ takes on a local minimum at y^*, then there must exist multipliers λ_i (real numbers) such that

$$\frac{\partial G(y^*)}{\partial y} = \sum_{i=1}^{n} \lambda_i \frac{\partial g_i(y^*)}{\partial y} = \lambda^T \frac{\partial g(y^*)}{\partial y} \tag{2.22}$$

where $\lambda = [\lambda_1, \ldots, \lambda_n]^T$.

Proof Since the rank of a matrix gives the maximum number of linearly independent rows, it follows from (2.21) that $\partial G(y^*)/\partial y$ must be a linear combination of the row vectors $\partial g_i(y^*)/\partial y$, $i \in N$. Equation (2.22) follows directly. ■

If y^* is a normal point of Y, then the matrix $\partial g(y^*)/\partial y$ is of maximum rank n (assuming $n<m$) and the multipliers are unique. If y^* is an abnormal point of Y, then $\partial g(y^*)/\partial y$ is less than maximum rank, say $r<n$, and r components of λ may be solved for in terms of the remaining $n-r$ components.

Multipliers such as λ play a central role in the optimization of parametric systems. They are generally referred to as *Lagrange* multipliers in deference to Comte Joseph Louis Lagrange, who utilized such variables in his classic work on analytical mechanics (Lagrange, 1788). Note that, by defining a new function

$$L(y,\lambda) \triangleq G(y)-\lambda^T g(y), \tag{2.23}$$

the necessary condition (2.22) reduces to

$$\left.\frac{\partial L}{\partial y}\right|_{y^*}=0. \tag{2.24}$$

Since any real number λ_i can be expressed as the difference

$$\lambda_i=\beta_i-\gamma_i$$

where $\beta_i \geqq 0$ and $\gamma_i \geqq 0$, it follows that the right-hand side of (2.22) defines a convex cone (the constraint cone, K) generated by the gradient vectors $\partial g(y^*)/\partial y$ and $-\partial g(y^*)/\partial y$. Thus (2.22) is equivalent to requiring that the vector $\partial G(y^*)/\partial y$ lie in the constraint cone K.

Example 2.11 Consider the problem of minimizing $G(\cdot)=y_1^2+y_2^2$, with the domain of $G(\cdot)$ defined by $g(\cdot)=y_2-y_1-1=0$. Figure 2.9 illustrates the relationship of the gradients at the minimum of $G(\cdot)$ on Y. For this two-dimensional case the constraint cone K at $y \in Y$ is a line perpendicular to $g(y)=0$. Only at y^* does $\partial G/\partial y$ lie in K. Forming

$$L=y_1^2+y_2^2-\lambda(y_2-y_1-1),$$

the analytical solution is obtained from (2.24) through the necessary

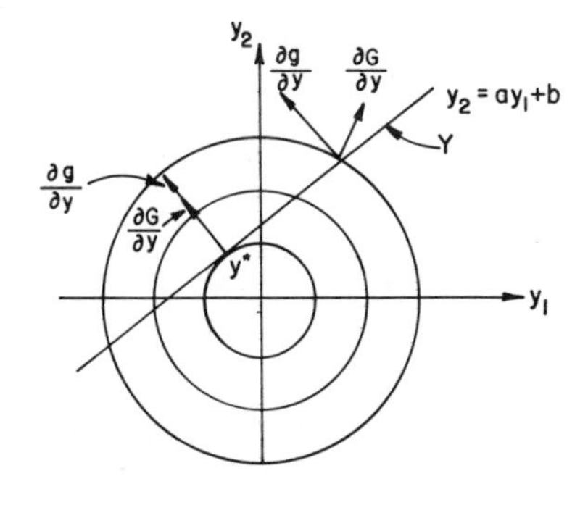

Figure 2.9. Geometry of cost and constraint gradients at optimal and nonoptimal points for Example 2.11.

conditions

$$\frac{\partial L}{\partial y_1} = 2y_1 + \lambda = 0$$

$$\frac{\partial L}{\partial y_2} = 2y_2 - \lambda = 0.$$

Eliminating λ between these equations and utilizing $g(\cdot) = 0$ yields $y_1 = -\frac{1}{2}$, $y_2 = \frac{1}{2}$.

2.5 AN ALTERNATE APPROACH FOR NORMAL INTERNAL POINTS

Since normality implies regularity, the latter is a weaker assumption and one that we use throughout. However, if we choose the stronger normality assumption instead, then an alternate derivation of first-order necessary conditions is possible.

Lemma 2.8 (Normal Internal Points) Let y^* be a normal internal point of Y. Let $G(\cdot): Y \to E^1$ and $g(\cdot): E^m \to E^n$ be C^1 in a neighborhood $B \cap Y$ of y^*. If $G(\cdot)$ takes on a local minimum at y^*, then the rank of the $(n\times 1)\times(m)$ matrix (assume $n<m$)

$$\begin{bmatrix} \dfrac{\partial G}{\partial y} \\ \dfrac{\partial g}{\partial y} \end{bmatrix}_{y^*} \tag{2.25}$$

is less than maximum rank.

Proof Assume that the rank of this matrix is $n+1$ at y^*. Then, for $\delta \in E^1$ with $\delta > 0$, the system of equations

$$\begin{bmatrix} \dfrac{\partial G}{\partial y} \\ \dfrac{\partial g}{\partial y} \end{bmatrix}_{y^*} e = \begin{bmatrix} -\delta \\ 0 \end{bmatrix}$$

would have a solution $e \in T$ defined by (2.10), which would contradict (2.19) and imply that y^* is not a local minimal point for $G(\cdot)$. ■

It follows from Lemma 2.8 that the rows of the matrix (2.25) must be linearly dependent at y^*. Thus there must exist real multipliers μ_0 and μ_i, $i \in N$, not all zero, such that

$$\mu_0 \frac{\partial G(y^*)}{\partial y} + \sum_{i=1}^{n} \mu_i \frac{\partial g_i(y^*)}{\partial y} = 0.$$

Since Lemma 2.8 requires that y^* be a normal point, rank $\partial g(y^*)/\partial y = n$ so that $\mu_0 \neq 0$. Dividing by μ_0 and letting $\lambda_i = -\mu_i/\mu_0$, we again obtain (2.22). Note that at any abnormal point $y^* \in Y$ the matrix in (2.25) is less than maximum rank. In other words, the condition of the normal internal point lemma is always satisfied at abnormal points, where the lemma itself is not applicable!

Example 2.12 Consider the problem of minimizing $G(\cdot) = y_1^2 + y_2^2 + y_3^2$ where the domain of $G(\cdot)$ is given by $Y = \{y \in E^3 \mid y_2 - y_1 - 1 = 0, (y_2 - y_1 - 1)^2 = 0\}$. Since every point of Y is abnormal, Lemma 2.8 is not applicable. However, Y is regular and the minimal point $(-\frac{1}{2}, \frac{1}{2}, 0)$ is obtained by employing condition (2.24).

2.6 SECOND-ORDER NECESSARY CONDITIONS FOR INTERNAL POINTS

Since $L(y, \lambda) = G(y)$ on all internal points of Y [where $g(y) = 0$ and $h(y) > 0$], it follows that local minimal points of $L(\cdot)$ on internal points of Y are also local minimal points of $G(\cdot)$ on internal points of Y and conversely. We use this observation in the proof of a second-order necessary condition given below.

It should be noted, however, that a fairly common misconception is that local minimal points of $L(\cdot)$ on E^m are local minimal points of $G(\cdot)$ on Y. This is the basis of some "multiplier rules" and is simply not true in general. Additional discussion is given by Vincent and Cliff (1970).

Lemma 2.9 Let y^* be a regular internal point of Y. Let $G(\cdot): Y \to E^1$ and $g(\cdot): E^m \to E^n$ be C^2 in a neighborhood $B \cap Y$ of y^* and let rank $\partial g(y^*)/\partial y$ be less than m. If $G(\cdot)$ takes on a local minimum at y^*, then the zeros of the $(m+n)\times(m+n)$ determinant

$$D(\Lambda) = \left| \begin{matrix} \dfrac{\partial^2 L}{\partial y^2} - \Lambda I & \left[\dfrac{\partial g}{\partial y}\right]^T \\ \dfrac{\partial g}{\partial y} & 0 \end{matrix} \right|_{y^*} \tag{2.26}$$

must be nonnegative.

Proof Since a minimal point for $L(\cdot)$ is also stationary at internal points of Y, Lemma 2.2 requires

$$Q = \tfrac{1}{2} e^T \left.\frac{\partial^2 L}{\partial y^2}\right|_{y^*} e \geqq 0 \tag{2.27}$$

for all $e \in T$ under the assumption that $L(\cdot)$ is C^2.

Consider the problem of determining e to yield the minimum of the (real) quadratic form

$$Q = \tfrac{1}{2} e^T \left.\frac{\partial^2 L}{\partial y^2}\right|_{y^*} e \tag{2.28}$$

subject to the constraint that e be a unit (i.e., nonzero) tangent vector in T at a regular internal point $y^* \in Y$. In other words, we constrain e to satisfy the following equations:

$$1 - e^T e = 0 \tag{2.29}$$

$$\frac{\partial g(y^*)}{\partial y} e = 0. \tag{2.30}$$

Such a minimum clearly exists since Q is a quadratic form on a vector space defined by the nonzero solutions to (2.30) normalized by (2.29). The existence of nonzero solutions for e defined by (2.30) is guaranteed by the assumption that rank $\partial g(y^*)/\partial y<m$.

If e minimizes (2.28) subject to (2.29) and (2.30), then condition (2.22) of Lemma 2.7 implies the existence of multipliers $\Lambda/2$ and $\eta=[\eta_1,\ldots,\eta_n]^T$ such that

$$\frac{\partial}{\partial e}\left[\tfrac{1}{2}e^T\frac{\partial^2 L}{\partial y^2}\bigg|_{y^*}e\right]+\frac{\Lambda}{2}\frac{\partial}{\partial e}[1-e^Te]+\eta^T\frac{\partial}{\partial e}\left[\frac{\partial g(y^*)}{\partial y}e\right]=0.$$

Performing the indicated differentiation, we get

$$e^T\left[\frac{\partial^2 L}{\partial y^2}\bigg|_{y^*}-\Lambda I\right]+\eta^T\frac{\partial g(y^*)}{\partial y}=0. \tag{2.31}$$

Transposing (2.31) and using (2.29) and (2.30), we have that the $m+n+1$ variables $e^T=[e_1,\ldots,e_m]$, Λ, and $\eta^T=[\eta_1,\ldots,\eta_n]$ must furnish a nonzero solution to the $m+n$ equations

$$\left[\frac{\partial^2 L}{\partial y^2}\bigg|_{y^*}-\Lambda I\right]e+\left[\frac{\partial g(y^*)}{\partial y}\right]^T\eta=0 \tag{2.32}$$

$$\frac{\partial g(y^*)}{\partial y}e=0 \tag{2.33}$$

with e being a unit vector. Thus the determinant given by (2.26) must be zero. In particular, Λ must be a root of the $(m-n)$ degree polynomial equation

$$D(\Lambda)=0. \tag{2.34}$$

Let Λ be any root of (2.34) and let e and η be the corresponding solutions to (2.32) and (2.33) with e being a unit vector. Post-multiplying (2.31) by e and invoking (2.33) yields

$$e^T\frac{\partial^2 L}{\partial y^2}\bigg|_{y^*}e=\Lambda. \tag{2.35}$$

It follows from (2.35) that the $(m-n)$ roots of (2.34) are real. Condition (2.27) requires that the roots be nonnegative. ■

The reader may wish to consult Hancock (1960) for a historical perspective on the development of Lemma 2.9.

2.7 NECESSARY CONDITIONS AND SUFFICIENT CONDITIONS AT REGULAR INTERNAL POINTS

We summarize the results obtained so far in the following theorem.

Theorem 2.1 Let y^* be a regular internal point of Y. Let $G(\cdot): Y \to E^1$ and $g(\cdot): E^m \to E^n$ be C^1 in a neighborhood of y^*. If $G(\cdot)$ takes on a local minimum at y^*, then multipliers $\lambda = [\lambda_1, \ldots, \lambda_n]^T$ exist such that

$$\left.\frac{\partial L}{\partial y}\right|_{y^*} = 0 \tag{2.36}$$

and

$$g(y^*) = 0 \tag{2.37}$$

where

$$L(y, \lambda) = G(y) - \lambda^T g(y). \tag{2.38}$$

In addition, if $G(\cdot)$ and $g(\cdot)$ are C^2 in a neighborhood of y^*, then the zeros of the determinant

$$D(\Lambda) = \left.\begin{vmatrix} \dfrac{\partial^2 L}{\partial y^2} - \Lambda I & \left[\dfrac{\partial g}{\partial y}\right]^T \\ \dfrac{\partial g}{\partial y} & [0] \end{vmatrix}\right|_{y^*} \tag{2.39}$$

must be nonnegative.

Proof The conditions follow directly from Lemma 2.7 and Lemma 2.9. ■

The conditions of Theorem 2.1 may be used to generate candidates for local internal minimal points. The second-order condition (2.39) can often be used to eliminate multiple candidates obtained from the first-order condition (2.36). However, due to the complexity of (2.39) the second-order condition would not be used in practice unless we were going to test for sufficiency as provided by the following theorem.

Theorem 2.2 Let y^* be a regular internal point of Y. Let $G(\cdot): Y \to E^1$ and $g(\cdot): E^m \to E^n$ be C^2 in a neighborhood of $y^* \in Y$.

The point y^* is a local minimal point for $G(\cdot)$ if

$$\left.\frac{\partial L}{\partial y}\right|_{y^*} = 0$$

$$g(y^*) = 0$$

and the zeros of $D(\Lambda)$ are all positive, where L is defined by (2.38) and $D(\Lambda)$ is defined by (2.39).

Proof To establish the theorem we show that y^* is a local minimal point for $L(\cdot)$ on internal points of Y. Since $L(\cdot) = G(\cdot)$ on internal points of Y, y^* is also a local minimal point for $G(\cdot)$ on internal points of Y.

By hypothesis, the equation $D(\Lambda) = 0$ is satisfied at y^* and all of its roots are positive. For each such Λ, there exists a vector $\eta \in E^n$ and a unit vector $e \in E^m$ such that (2.32) and (2.33) are satisfied. [Note that (2.33) defines all tangent vectors.] Using the same device as before, we post-multiply (2.31), the transpose of (2.32), by e to get

$$e^T \left.\frac{\partial^2 L}{\partial y^2}\right|_{y^*} e = \Lambda > 0.$$

Let $\delta y(\cdot): (0, \gamma) \to E^m$ generate a unit vector e tangent to Y at y^*. From the second-order approximation theorem we have

$$L[y^* + \alpha \delta y(\alpha)] - L(y^*) = \alpha^2 \left\{ \tfrac{1}{2} e^T \left.\frac{\partial^2 L}{\partial y^2}\right|_{y^*} e + \frac{R(\alpha^2)}{\alpha^2} \right\}$$

where $R(\alpha^2)/\alpha^2 \to 0$ as $\alpha \to 0$, which implies that y^* is a minimal point for $L(\cdot)$ and therefore $G(\cdot)$. ■

We note that, if $Y=E^m$ (no equality constraints), then the zeros of $D(\Lambda)$ are just the eigenvalues of $\partial^2 G/\partial y^2$ and we arrive at the well known result that the definiteness of a quadratic form may be determined by examining its eigenvalues.

Example 2.13 Determine all local minima for the function $G(\cdot)=y_1^2+y_2^2$ with its domain Y defined by all $y\in E^2$ such that $g_1(y)=2y_1^2+y_2-8=0$. We obtain the solution by first forming

$$L=y_1^2+y_2^2-\lambda(2y_1^2+y_2-8)$$

so that the first-order necessary conditions are given by

$$\frac{\partial L}{\partial y_1}=2y_1-4\lambda y_1=0$$

$$\frac{\partial L}{\partial y_2}=2y_2-\lambda=0,$$

which yield the candidate solutions

$$y_1=0, \qquad y_2=8, \quad \lambda=16 \tag{2.40}$$

$$y_1=+\sqrt{\tfrac{31}{8}}, \quad y_2=\tfrac{1}{4} \quad \lambda=\tfrac{1}{2} \tag{2.41}$$

$$y_1=-\sqrt{\tfrac{31}{8}}, \quad y_2=\tfrac{1}{4}, \quad \lambda=\tfrac{1}{2}. \tag{2.42}$$

Since the first-order necessary conditions have produced multiple candidates, we employ the second-order conditions to eliminate erroneous candidates. By taking second partial derivatives of L we form

$$D(\Lambda)=\begin{vmatrix} 2-4\lambda-\Lambda & 0 & 4y_1 \\ 0 & 2-\Lambda & 1 \\ 4y_1 & 1 & 0 \end{vmatrix}.$$

The roots of $D(\Lambda)$ are obtained by setting the above determinant equal to zero, yielding

$$\Lambda=\frac{2-4\lambda+32y_1^2}{1+16y_1^2}.$$

Evaluating Λ for each of the candidate solutions yields

$$\Lambda_1 = -62$$

$$\Lambda_2 = \tfrac{124}{63}$$

$$\Lambda_3 = \tfrac{124}{63}.$$

Thus solutions (2.41) and (2.42) satisfy the second-order sufficient condition and are local minimal points. The second-order sufficient condition fails for (2.40). It would remain as a candidate for a minimum except for the fact that a negative value for Λ_1 is sufficient for (2.40) to be a local maximal point.

2.8 FIRST-ORDER NECESSARY CONDITIONS FOR MINIMIZING ON *Y*

In general, Q^* is not empty (i.e., some of the inequality constraints are active) and we must satisfy the necessary condition of Lemma 2.1:

$$\frac{\partial G(y^*)}{\partial y} e \geqq 0 \tag{2.43}$$

for all $e \in T$ where now, for a regular set Y, the tangent cone T at y^* is defined by all e that satisfy

$$\frac{\partial g(y^*)}{\partial y} e = 0 \tag{2.44}$$

and

$$\frac{\partial \hat{h}(y^*)}{\partial y} e \geqq 0. \tag{2.45}$$

Here $\hat{h}(y^*) = [h_1(y^*), \ldots, h_{q^*}(y^*)]^T$ denotes the active inequality constraints at y^*.

If we think of (2.44) as being representable in terms of two inequality conditions

$$\frac{\partial g(y^*)}{\partial y} e \geqq 0$$

$$-\frac{\partial g(y^*)}{\partial y} e \geqq 0.$$

then by Lemma 2.5 we have that T is precisely the polar to the constraint cone K generated by the vectors $\partial h_j(y^*)/\partial y$, $j \in Q^* = \{1,\ldots,q^*\}$, $\partial g_i(y^*)/\partial y$, $i \in N = \{1,\ldots,n\}$, and $-\partial g_i(y^*)/\partial y$, $i \in N$.

Lemma 2.10 Let y^* be a regular point of Y. Let $G(\cdot): Y \to E^1$, $g(\cdot): E^m \to E^n$, and $\hat{h}(\cdot): E^m \to E^{q^*}$ be C^1 in a neighborhood $B \cap Y$ of y^*. If $G(\cdot)$ takes on a local minimum at y^*, then there must exist multipliers λ_i and $\hat{\mu}_j$ such that

$$\frac{\partial G(y^*)}{\partial y} = \sum_{i=1}^{n} \lambda_i \frac{\partial g_i(y^*)}{\partial y} + \sum_{j=1}^{q^*} \hat{\mu}_j \frac{\partial \hat{h}_j(y^*)}{\partial y} \tag{2.46}$$

and

$$\hat{\mu}_j \geqq 0 \qquad \forall j \in Q^* = \{1,\ldots,q^*\}. \tag{2.47}$$

Proof Since $T = K^*$, it follows from Theorem 1.7 (Farkas' lemma) that the vector $\partial G(y^*)/\partial y$ in (2.43) is an element of the constraint cone K. Thus there must exist *nonnegative* multipliers $\hat{\mu}_j$, $j \in Q^*$, β_i, $i \in N$, and γ_i, $i \in N$, such that

$$\frac{\partial G(y^*)}{\partial y} = \sum_{j=1}^{q^*} \hat{\mu}_j \frac{\partial \hat{h}_j(y^*)}{\partial y} + \sum_{i=1}^{n} \beta_i \frac{\partial g_i(y^*)}{\partial y} - \sum_{i=1}^{n} \gamma_i \frac{\partial g_i(y^*)}{\partial y}.$$

By letting $\lambda_i = \beta_i - \gamma_i$ $\forall i \in N$, the proof is complete. ■

Note that the multipliers λ_i may be positive, negative, or zero. It follows that (2.46) implies the condition

$$\operatorname{rank}\begin{bmatrix} \dfrac{\partial G}{\partial y} \\[2mm] \dfrac{\partial g}{\partial y} \\[2mm] \dfrac{\partial \hat{h}}{\partial y} \end{bmatrix}_{y^*} = \operatorname{rank}\begin{bmatrix} \dfrac{\partial g}{\partial y} \\[2mm] \dfrac{\partial \hat{h}}{\partial y} \end{bmatrix}_{y^*}.$$

However, the converse is not generally true due to the requirement that $\hat{\mu}_j \geqq 0$. Let

$$\lambda = [\lambda_1,\ldots,\lambda_n]^T, \qquad \mu = [\mu_1,\ldots,\mu_q]^T$$

and define

$$L(y, \lambda, \mu)=G(y)-\lambda^T g(y)-\mu^T h(y). \tag{2.48}$$

It follows from (2.46) and (2.47) that the first-order necessary conditions for a local minimum of $G(\cdot)$ at a regular point $y^* \in Y$ may be written equivalently as

$$\left.\frac{\partial L}{\partial y}\right|_{y^*}=0 \tag{2.49}$$

$$\mu^T h(y^*)=0 \tag{2.50}$$

$$\mu \geqq 0. \tag{2.51}$$

Formulas (2.50) and (2.51) imply that $\mu_j=0$ if $h_j(h^*)>0$, that is, if $j \notin Q^*$. In other words, $\mu_j=0$ if $h_j(\cdot)$ is not an active constraint at y^*.

2.9 SECOND-ORDER NECESSARY CONDITIONS FOR MINIMIZING ON *Y*

In general $L(y, \lambda, \mu) \neq G(y)$ on Y and we are unable to proceed directly to second-order conditions in terms of $L(\cdot)$ as we did in Section 2.6 for internal minimal points. We do note, however, that if y^* minimizes $L(y, \lambda, \mu)$ on Y, then y^* minimizes $G(y)$ on Y. To verify this, choose $y \in Y$, with y in a sufficiently small ball about y^* such that the inactive constraints $h_j(\cdot)$, $j \notin Q^*$, remain inactive. Such a ball exists since the $h_j(\cdot)$ are continuous with $h_j(y^*)>0 \; \forall j \notin Q^*$.

If y^* minimizes $L(\cdot)$, we have

$$G(y)-\lambda^T g(y)-\sum_{j=1}^{q^*} \mu_j h_j(y) \geqq G(y^*)-\lambda^T g(y^*)-\sum_{j=1}^{q^*} \mu_j h_j(y^*).$$

Since $g(y)=0 \; \forall y \in Y$ and $h_j(y^*)=0 \; \forall j \in Q^*$, the above equation becomes

$$G(y)-\sum_{j=1}^{q^*} \mu_j h_j(y) \geqq G(y^*).$$

Noting that $\mu_j \geqq 0$ and $h_j(y) \geqq 0 \; \forall j \in Q^*$, we have

$$G(y) \geqq G(y^*) + \sum_{j=1}^{q^*} \mu_j h_j(y) \geqq G(y^*),$$

which implies that y^* also minimizes $G(y)$.

Thus minimizing $L(\cdot)$ on Y minimizes $G(\cdot)$ on Y. The converse is not true in general because $h_j(y^*)=0$ does not imply $h_j(y)=0$ in a ball about y^*. If an assumption is made that there exists an arc $y^*+\alpha\,\delta y(\alpha) \in Y$ such that both $g[y^*+\alpha\,\delta y(\alpha)]=0$ and $h_j[y^*+\alpha\,\delta y(\alpha)]=0 \; \forall j \in Q^*$, then $L(\cdot)=G(\cdot)$ on this arc. In such a case minimizing $G(\cdot)$ on Y implies minimizing $L(\cdot)$ on Y.

In order to develop second-order necessary conditions similar to (2.26), we need a situation in which minimizing $G(\cdot)$ on Y implies minimizing $L(\cdot)$ on Y. We employ the following second-order constraint qualification introduced by Fiacco and McCormick (1968).

DEFINITION 2.9 A point $y^* \in Y$ is said to satisfy a *second-order constraint qualification* if and only if the following conditions are satisfied, where $\hat{h}(\cdot)=[h_1(\cdot),\ldots,h_{q^*}(\cdot)]^T$ denotes the active inequality constraints:

(i) $g(y^*)=0$; (2.52)

(ii) $\hat{h}(y^*)=0$; (2.53)

(iii) $G(\cdot)$, $g(\cdot)$, and $\hat{h}(\cdot)$ are C^2 on $B \cap Y$ where B is some ball centered at y^*;

(iv) there exists a nonzero $e \in E^m$ satisfying

$$\frac{\partial g(y^*)}{\partial y} e = 0 \tag{2.54}$$

and

$$\frac{\partial \hat{h}(y^*)}{\partial y} e = 0 \tag{2.55}$$

such that e is tangent to a C^2 arc $y^*+\alpha\,\delta y(\alpha) \in Y$, $\alpha \in [0,\gamma]$ [i.e., $\delta y(\alpha) \to e$ as $\alpha \to 0$] where

$$g[y^*+\alpha\,\delta y(\alpha)]=0 \qquad \forall \alpha \in [0,\gamma]$$

$$h[y^*+\alpha\,\delta y(\alpha)]=0 \qquad \forall \alpha \in [0,\gamma].$$

If the second-order constraint qualification is satisfied at a regular point $y^* \in Y$ that minimizes $G(\cdot)$ on Y, then $L(y,\lambda,\mu)=G(y)$ along the arc $y^*+\alpha\,\delta y(\alpha)$. The same arguments employed in Section 2.6 yield the following additional (second-order) necessary conditions for y^* to be a minimal point:

(i)

$$e^T \left.\frac{\partial^2 L}{\partial y^2}\right|_{y^*} e \geqq 0 \tag{2.56}$$

for all e satisfying (2.54) and (2.55);

(ii) the $m-(n+q^*)$ zeros of the $(m+n+q^*)$-order determinant

$$D(\Lambda)=\left|\begin{matrix} \dfrac{\partial^2 L}{\partial y^2}-\Lambda I & \left[\dfrac{\partial g}{\partial y}\right]^T & \left[\dfrac{\partial \hat{h}}{\partial y}\right]^T \\ \dfrac{\partial g}{\partial y} & [0] & [0] \\ \dfrac{\partial \hat{h}}{\partial y} & [0] & [0] \end{matrix}\right|_{y^*} \tag{2.57}$$

must be nonnegative, where $\hat{h}(\cdot)=[h_1(\cdot),\ldots,h_{q^*}(\cdot)]^T$ is composed of the active inequality constraints at y^*.

Similarly, the arguments of Section 2.7 imply that if the zeros of $D(\Lambda)$ are all strictly positive at a regular point y^* for which the second-order constraint qualification is satisfied, then y^* is a local minimal point for $G(\cdot)$ on Y.

Due to the severity of conditions (2.54) and (2.55) (compared with the regularity conditions) there may not exist a nonzero e that satisfies the second-order constraint qualification. This can happen, for example, at a "corner" of Y where every nonzero admissible direction of motion (i.e., tangent vector) would cause one or more active inequality constraints to become inactive.

2.10 NECESSARY CONDITIONS AND SUFFICIENT CONDITIONS FOR MINIMIZING ON Y

We summarize the results of Sections 2.8 and 2.9 in the following theorems.

Theorem 2.3 Let y^* be a regular point of Y. Let $G(\cdot): Y \to E^1$, $g(\cdot): E^m \to E^n$, and $h(\cdot): E^m \to E^q$ be C^1 in a neighborhood of y^*. If $G(\cdot)$ takes on a local minimum at y^*, then multipliers $\lambda=[\lambda_1,\ldots,\lambda_n]^T$ and $\mu=[\mu_1,\ldots,\mu_q]^T$ exist such that

$$\left.\frac{\partial L}{\partial y}\right|_{y^*}=0 \tag{2.58}$$

$$g(y^*)=0 \tag{2.59}$$

$$h(y^*)\geqq 0 \tag{2.60}$$

$$\mu^T h(y^*)=0 \tag{2.61}$$

$$\mu\geqq 0 \tag{2.62}$$

where

$$L(y,\lambda,\mu)\triangleq G(y)-\lambda^T g(y)-\mu^T h(y). \tag{2.63}$$

In addition, if $G(\cdot)$, $g(\cdot)$, and $\hat{h}(\cdot)$ are C^2 in a ball about y^* and if the second-order constraint qualification holds, then the $m-(n+q^*)$ zeros of the $(m+n+q^*)$-order determinant

$$D(\Lambda)\triangleq\left|\begin{matrix} \dfrac{\partial^2 L}{\partial y^2}-\Lambda I & \left[\dfrac{\partial g}{\partial y}\right]^T & \left[\dfrac{\partial \hat{h}}{\partial y}\right]^T \\ \dfrac{\partial g}{\partial y} & [0] & [0] \\ \dfrac{\partial \hat{h}}{\partial y} & [0] & [0] \end{matrix}\right|_{y^*} \tag{2.64}$$

must be nonnegative.

Proof The theorem follows directly from Lemma 2.10 and the results of Section 2.9. ■

Again, the second-order necessary condition would rarely be used unless we were going to test for sufficiency as provided by the following theorem.

Theorem 2.4 Let y^* be a regular point of Y. Let $G(\cdot): Y\rightarrow E^1$, $g(\cdot): E^m\rightarrow E^n$, and $h(\cdot): E^m\rightarrow E^q$ be C^2 in a neighborhood of y^*. If the second-order constraint qualification holds at y^* for a nonzero vector e, then y^* is a local minimum point for $G(\cdot)$ if conditions (2.58)-(2.62) hold and the zeros of

$D(\Lambda)$ are all positive, where $L(y, \lambda, \mu)$ and $D(\Lambda)$ are defined by (2.63) and (2.64), respectively.

Proof The proof of this theorem follows the proof of Theorem 2.2 with the additional constraints that $\hat{h}(y)=0$ along the arc defined in the second-order constraint qualification. ■

Example 2.14 Determine all local minima for the function $G(\cdot)=-y_1^2-y_2^2$ with its domain Y defined by $Y=\{y\in E^2 | y_1y_2-1=0,\ y_1\geqq 0,\ y_2\geqq 0,\ 2-y_1\geqq 0,\ 2-y_2\geqq 0\}$. We first form

$$L=-y_1^2-y_2^2-\lambda(y_1y_2-1)-\mu_1y_1-\mu_2y_2-\mu_3(2-y_1)-\mu_4(2-y_2)$$

so that the first-order necessary conditions are given by

$$\frac{\partial L}{\partial y_1}=-2y_1-\lambda y_2-\mu_1+\mu_3=0$$

$$\frac{\partial L}{\partial y_2}=-2y_2-\lambda y_1-\mu_2+\mu_4=0.$$

We next check for an internal solution by setting $\mu_1=\mu_2=\mu_3=\mu_4=0$ (all inequality constraints inactive). Eliminating λ yields the condition

$$y_2=\pm y_1.$$

The negative solution is not in Y and is discarded. Employing the equality conditions results in the candidate

$$y_1=y_2=1.$$

Other candidates are obtained by examining the necessary conditions with one or more of the inequality constraints active. For example, if only h_3 is active, so that $y_1=2(\mu_3\geqq 0)$, $\mu_1=\mu_2=\mu_4=0$, then $y_2=\frac{1}{2}$ from the equality constraint and the necessary conditions become

$$-4-\frac{\lambda}{2}+\mu_3=0$$

$$-1-2\lambda=0.$$

Clearly $\lambda=-\frac{1}{2}$ and the first necessary condition is satisfied by $\mu_3=\frac{15}{4}>0$ so that $(2,\frac{1}{2})$ is a candidate. Similarly, if only h_1 is active, so that

$y_1=0(\mu_1 \geqq 0)$, $\mu_2=\mu_3=\mu_4=0$, then $y_2=\infty$ from the equality constraint. However, this point is not in Y and hence not a candidate. In general, all combinations of active equality constraints must be examined. For this problem, there is one more candidate at $(\frac{1}{2},2)$.

The second-order condition for the internal candidate is given by

$$D(\Lambda)=\begin{vmatrix} -2-\Lambda & -\lambda & y_2 \\ -\lambda & -2-\Lambda & y_1 \\ y_2 & y_1 & 0 \end{vmatrix}$$

with $y_1=y_2=1$ and $\lambda=-2$. Setting $D(\Lambda)=0$ yields $\Lambda=-4$, which is sufficient for a local maximal point.

The second-order condition is not applicable at the other two candidate points since the second-order constraint qualification is not satisfied. At both points the only e vector that satisfies both (2.54) and (2.55) is the zero vector. Since the theorem of Weierstrass is satisfied for this problem, a minimum exists and must be at one or both of the remaining candidates. Direct calculation of the cost function yields the same cost $(G=-\frac{15}{4})$ at both candidate points. Hence both are global minimal points.

Recall that internal local minimal points are also stationary. According to Theorem 2.3, if a local minimal point is not internal, then it may $(\mu=0)$ or may not $(\mu\neq 0)$ be stationary. Usually, however, as in the previous example, noninternal points are not stationary. This implies that the principle of flat laxity is generally not applicable at noninternal minimal points, and at such points the value of the argument of the cost function may indeed be critical.

Since Theorem 2.3 is valid for all $y^* \in Y$, it is valid for all internal points. Thus Theorem 2.3 includes Theorem 2.1 as a special case. Because of this, we often use Theorem 2.1 first to seek internal minimal point candidates and then impose the inequality constraints, usually one at a time or in combinations, to seek the noninternal candidates. The following observations, which follow directly from Theorem 2.3, are often useful when using this procedure:

1 Any candidate solution (internal or not) remains as a candidate solution when one or more new inequality constraints are imposed if and only if the original candidate solutions satisfy the newly imposed inequality conditions.

2 Imposing one or more new inequality constraints does not introduce any new candidates that lie off of the set defined by the new active inequality constraints. Any new candidates must lie on this set.

2.11 LINEAR SYSTEMS

There is a large class of parametric optimization problems for which the cost function and the constraint equations are linear. For example, a linear cost function may be represented by

$$G(y)=c^T y \tag{2.65}$$

with linear constraint equations defining Y given by

$$g(y)=Ay-a=0 \tag{2.66}$$

$$h(y)=By-b\geqq 0 \tag{2.67}$$

where c^T is a constant $1\times m$ row vector, A is a constant $n\times m$ matrix, B is a constant $q\times m$ matrix, a is an $n\times 1$ column vector, and b is a $q\times 1$ column vector. The determination of $y\in Y$ to minimize G for this case is often referred to as a *linear programming problem* (Hadley, 1962). "Programming" is used in deference to the fact that algorithms for use on the digital computer are available for directly generating solutions even when m is quite large. By contrast, the optimization problem of the previous section is often referred to as a general programming or *nonlinear programming problem*. Algorithms for use on the digital computer are also available for solving nonlinear problems; however, they are usually not so simple to implement and are often successful for only a limited class of nonlinear problems.

For linear systems the first-order approximation theorem is exact; that is, the remainder term in (1.15) is zero. It then follows, using arguments similar to the proof of Lemma 2.1, that

$$\frac{\partial G(y^*)}{\partial y}e=c^T e\geqq 0 \tag{2.68}$$

$\forall e\in T$ is both necessary *and* sufficient for y^* to be a local minimal point for $G(\cdot)$ on Y. Furthermore, as in Section 2.2, the first-order approximation theorem (which is now exact) provides

$$\frac{\partial g(y^*)}{\partial y}e = Ae = 0 \tag{2.69}$$

$$\frac{\partial \hat{h}(y^*)}{\partial y}e = \hat{B}e \geqq 0 \tag{2.70}$$

for every vector tangent to Y at y^* where $\hat{B}$ corresponds to the $q \times m$ matrix of coefficients for the active inequality constraints.

Lemma 2.11 If the constraint set Y is defined by the linear relations (2.66) and (2.67), then Y is a regular set.

Proof Let T be the tangent cone to Y at a point $y \in Y$. Then $e \in T$ if and only if there exists a positive number γ and a continuous function $\delta y(\cdot):[0,\gamma] \to E^m$ such that $\delta y(\alpha) \to e$ as $\alpha \to 0$ and for all $\alpha \in [0,\gamma]$

$$A\left[y + \alpha\,\delta y(\alpha)\right] = a \tag{2.71}$$

$$B\left[y + \alpha\,\delta y(\alpha)\right] \geqq b. \tag{2.72}$$

From (2.66) and (2.71) we have

$$\alpha A\,\delta y(\alpha) = 0.$$

Dividing by $\alpha > 0$ and taking the limit as $\alpha \to 0$ yields

$$Ae = 0 \tag{2.73}$$

For $j \in Q = \{1, \ldots, q\}$ let B^j denote row j of matrix B, let b_j denote element j of vector b, and define

$$Q^* \triangleq \left\{ j \in Q \mid B^j y = b_j \right\}.$$

Then from (2.72), for any $\delta y(\cdot)$,

$$B^j\left[y + \alpha\,\delta y(\alpha)\right] > b_j \tag{2.74}$$

for all $j \notin Q^*$ and for all sufficiently small $\alpha > 0$. Thus (2.72) yields

$$\alpha B^j\,\delta y(\alpha) \geqq b_j - B^j y = 0$$

for all $j \in Q^*$. Dividing by $\alpha > 0$ and taking the limit as $\alpha \to 0$ yields

$$B^j e \geqq 0 \tag{2.75}$$

for all $j \in Q^*$.

Thus every vector $e \in T$ must satisfy (2.73) and (2.75). To establish that Y is a regular set we must show that at each $y \in Y$ every vector e satisfying (2.73) and (2.75) is in fact tangent to Y.

Let y and e be such that (2.66), (2.67), (2.73), and (2.75) are satisfied. Define $\delta y(\alpha) = e$. From (2.66) and (2.73) we have

$$A[y + \alpha e] = a$$

for all α. From (2.74) we have

$$B^j[y + \alpha e] > b_j$$

for all $j \notin Q^*$ and for all sufficiently small α. From (2.67) and (2.75) we have

$$B^j[y + \alpha e] = b_j + \alpha B^j e \geqq b_j$$

for all $j \in Q^*$ and for all $\alpha \geqq 0$. Hence $e \in T$. ■

From Lemma 2.11 the constraint set Y is regular so that the tangent cone T to Y at y^* consists of exactly those vectors e satisfying (2.69)–(2.70). Writing the equality condition (2.69) as two inequality conditions, it follows from Farkas' lemma that there must exist multipliers $\lambda = [\lambda_1, \dots, \lambda_n]^T$, $\hat{\mu} = [\hat{\mu}_1, \dots, \hat{\mu}_q]^T$ such that

$$c^T = \lambda^T A + \hat{\mu}^T \hat{B} \tag{2.76}$$

with $\hat{\mu} \geqq 0$. We summarize these results with the following theorem.

Theorem 2.5 Let y^* be a point of $Y = \{y \in E^m \mid Ay = a \text{ and } By \geqq b\}$ and let $G = c^T y$. Then $G(\cdot)$ takes on a minimum at y^* if and only if there exist multipliers $\lambda = [\lambda_1, \dots, \lambda_n]^T$ and $\mu = [\mu_1, \dots, \mu_q]^T$ such that

$$c^T = \lambda^T A + \mu^T B \tag{2.77}$$

$$Ay = a \tag{2.78}$$

$$By \geqq b \tag{2.79}$$

$$\mu^T(By - b) = 0 \tag{2.80}$$

$$\mu \geqq 0. \tag{2.81}$$

Proof The theorem follows directly from conditions (2.68)–(2.70) and (2.76). ■

Note that condition (2.77) requires that the vector c lie in the constraint cone K. In this case the cone K is generated by the row vectors of the matrices A and B. We also note from (2.65), (2.66), (2.77) and (2.80) that the minimum cost is given by

$$G = \lambda^T a + \mu^T b,$$

which demonstrates the multipliers to be sensitivity coefficients (see also Section 4.3).

Since (2.77) is independent of y for all internal points of Y (i.e., $By < b$, $\mu \equiv 0$), it follows that, if any one internal point of Y satisfies (2.77), then all internal points must satisfy (2.77) and in addition all such points must have the same cost $G = \lambda^T a$ since the minimum cost is unique. This is possible, but rather uninteresting! Note that when the internal points are minimal, then the same minimal cost is also achieved on at least one of the active inequality constraints since $\mu \equiv 0$ is a possible option with an inequality constraint active. Thus for linear systems the minimal cost can always be achieved at points that lie on at least one active inequality constraint.

Example 2.15 Determine all minimal points for the function $G(\cdot) = y_1 + y_2$ with its domain defined by $Y = \{y \in E^2 | -y_1 + y_2 - 1 = 0,\ 2 - y_1 \geqq 0,\ y_2 - 2 \geqq 0\}$. For this problem we have

$$c^T = [1, 1]$$

$$A = [-1, 1]$$

$$a = 1$$

$$B = \begin{bmatrix} -1 & 0 \\ 0 & 1 \end{bmatrix}$$

$$b^T = [-2, 2].$$

Condition (2.77) yields

$$[1,1]=\lambda[-1,1]+[\mu_1,\mu_2]\begin{bmatrix}-1 & 0\\ 0 & 1\end{bmatrix}$$

or equivalently

$$1=-\lambda-\mu_1$$

$$1=\lambda+\mu_2,$$

which is satisfied by $\mu_1=0$, $\lambda=-1$, $\mu_2=2$ at the minimal point $y_1=1$, $y_2=2$ with cost $G=3$.

Note that, if the equality constraint were given by $y_1+y_2=0$, then all points satisfying $y_1+y_2=0$ for $-2\leqq y_1\leqq 2$ would be minimal, yielding the same minimal cost $G=0$.

2.12 EXERCISES

2.1 The intrinsic growth rate I for a given species of plant has been proposed to be given by

$$I=\frac{FN}{e^{-1}+N}$$

where N is soil nutrient, e is an efficiency factor, and F is a growth factor. Experiments show F and e to be related as shown in the figure. Using the qualitative relation that $F=f(e^{-1})$, show that e^*, the value of e that maximizes I, can be obtained by drawing a tangent to the curve starting from $-N$ as shown in the figure below.

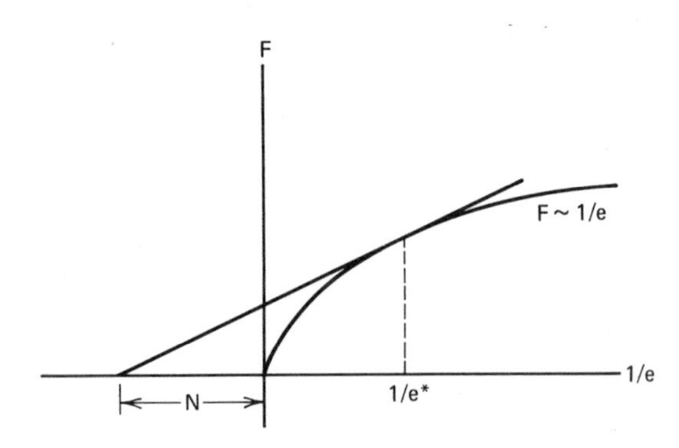

2.2 A commonly used 16 oz beer can is 6.25 in. in height and 2.5 in. in diameter. Determine the percent waste in material that results from using this nonoptimal combination of dimensions.

2.3 Consider the problem of minimizing the scalar function $G(y) = -(y_1^2 + y_2^2)$ subject to the constraint $g(y) = (y_1 - 1)^2 + y_2^2 - 4 = 0$. Find the points y for which

$$\frac{\partial G}{\partial y}\delta y > 0$$

$\forall y + \delta y \in Y = \{y \mid g(y) = 0\}$. Discuss why this condition is not sufficient to insure a local minimum for $G(\cdot)$.

2.4 Using first-and second-order conditions, find the local minimum for

$$G = y_1^2 + \frac{\sqrt{3}}{2} y_2$$

subject to

$$g = y_1 - \sin y_2 = 0$$

with $0 < y_2 < \pi$. Use both Lagrange multipliers and direct substitution (e.g., solving $g = 0$ for y_1 and substituting into G) to obtain solutions.

2.5 Determine all local minima for the function $G(y) = 8y_1^2 + 10y_2^2$ where $y \in Y = \{y \in E^2 \mid 3 - 4y_1 - 6y_2 = 0\}$.

2.6 Determine candidates for a local minimum for the function $G(y) = \sin y_1 + y_2^2$ where $y \in Y = \{y \in E^3 \mid y_2^2 + (y_3 - a)^2 = a^2$ and $y_2^2 + (y_3 + a)^2 = a^2\}$ and a is a constant.

2.7 Find the minimum distance between the origin and the line that satisfies the equations $y_1 + 2y_2 + 3y_3 = 10$ and $y_1 - y_2 + 3y_3 = 1$.

2.8 The lifeguard has been eying the girl in the bikini. Suddenly, she cries for help in the water at point A. He is on the sand at point B. Both of them are equal distances from the shore line. Consider B to be the origin of a coordinate system with the y_1 axis parallel to the shoreline. The coordinates of A are given by $(b, 2a)$. The lifeguard can run in the sand with speed V_s and swim in the water with speed V_w. Determine a relationship between his heading angles in the sand and water to minimize the time of rescue.

2.9 Let $Ay=b$ represent a system of linear equations where A is an $n\times m$ real constant matrix, y is an $m\times 1$ vector (unknown), and b is an $n\times 1$ real vector (constant). Assume that $n<m$ and that $r[A]=r[A \,\vdots\, b]=n$ so that solutions exist for y. Find the solution for y that minimizes $\|y\|^2=y^Ty$.

2.10 Given $G=y_1^2+y_2^2$ where $y\in Y=\{y\in E^2 \mid y_1y_2=1,\ 0\leqq y_i\leqq 2,\ i=1,2\}$. Does a minimum and maximum for $G(\cdot)$ exist on Y? Using Theorem 2.3, find all candidates for a local minimum and a local maximum for $G(\cdot)$. Can Theorem 2.4 [or condition (2.64) of Theorem 2.3] be used for this problem?

2.11 Find all points that are candidates for a local minimum for $G(\cdot)=-(y_1^2+y_2^2)$ where $y\in Y=\{y\in E^2 \mid 8-2y_1^2-y_2\geqq 0\}$. Check each candidate using the sufficiency condition of Theorem 2.4.

2.12 A terrain-following cruise missile flies toward its target over hilly terrain at a speed V (ft/sec) and a commanded clearance altitude h (ft). The lower and faster the missile flies, the higher the probability P_c is that it will "clobber" the ground. The higher the missile flies, the higher the probability P_d is that it will be detected and shot down. Assuming

$$P_d=1-e^{-h/200}$$

and

$$P_c=e^{-10h/V},$$

determine V and h to maximize the probability of success

$$P_s=(1-P_d)(1-P_c)$$

subject to the constraints

$$200\leqq V\leqq 300.$$

2.13 An implicit assumption made by many theoretical biologists is that natural selection results in animal behavior that maximizes individual "fitness." Using this assumption, Pulliam (1974) developed a feeding theory for the prediction of predator response to alternative prey types. The following problem is an application of Pulliam's theory. While searching for food, a fox encounters r rabbits per unit time and m field mice per unit time. He may choose to eat his prey or not at each encounter. Let y_r = fraction of rabbits eaten on each encounter and y_m = fraction of mice eaten on each encounter. The food value of rabbits is f_r per rabbit and the food value of mice is f_m per mouse. The fox must spend time searching for prey. Once found, he must spend time pursuing and handling. The search time t_s between encounters is assumed to be constant. The pursuing and handling time per rabbit is t_r and the pursuing and handling time for each mouse is t_m. Suppose the fox has spent t_s units of time searching. The total food value obtained will be

$$F = f_r r y_r t_s + f_m m y_m t_s$$

and the total time spent obtaining this food will be

$$T = t_r r y_r t_s + t_m m y_m t_s + t_s.$$

Thus the food value per unit time is

$$\frac{F}{T} = \frac{f_r x_r + f_m x_m}{1 + t_r x_r + t_m x_m}$$

where $x_r = r y_r$ and $x_m = m y_m$. Determine the optimal diet for the fox, that is, how should he choose y_r and y_m to maximize F/T? Note that $0 \leqq x_r \leqq r$, $0 \leqq x_m \leqq m$. Laboratory data show that $f_r/t_r > f_m/t_m$.

2.14 Find all candidates for a local minimum for the function $G(\cdot) = y_1^2 + 2y_2^2 + y_3^2$ where $y \in Y = \{y \in E^3 | 3y_2 + y_3 - 13 \geqq 0,\ 2y_1 + 3y_2 + y_3 - 19 = 0\}$.

2.15 Find all the points that are candidates for a local minimum for the function $G(\cdot) = 3(y_1+1)^2 + y_2$ where $y \in Y = \{y \in E^2 | 2y_1^2 + y_2 - 8 = 0,\ 2 - y_1 \geqq 0,\ y_2 \geqq 0,\ 8 - y_2 \geqq 0,\ 8 + 2y_1 - y_2 \geqq 0\}$.

2.16 Volcanic ash from Mt. St. Helens is to be mailed in a right circular cylindrical package with closed ends. Let r and h denote the radius and height of the package, respectively. Because of materials on hand, the surface area of the package is limited to 5000 in.^{2} or less. Furthermore, the Post Office requires that the sum of the largest dimension (termed the "length") plus the girth (circumference perpendicular to the "length") must be less than or equal to 84 in. Determine r and h for maximum volume. Which dimension (height or diameter) is the "length"?

2.17 Water is pumped from a tank, through a solar collector, and then back into the tank. Equilibrium conditions for a particular system on a certain day are modeled, approximately, by the equations

$$0 = 200 - 8(T_1 - 290) - v(T_1 - T_2)$$
$$0 = v(T_1 - T_2) - 2(T_2 - 290),$$

where T_1 is the temperature (°K) of the solar collector plate, T_2 is the temperature (°K) of the water in the tank, and v is proportional to to the flow rate, with

$$0 \leqq v \leqq 4.$$

Determine T_1 and v to maximize T_2.

2.18 A certain automobile shock absorber produces a force F given by the nonlinear relation

$$F(v) = 10v|v| = \begin{cases} 10\,v^2 & \text{if} \quad v \geqq 0 \\ -10\,v^2 & \text{if} \quad v \leqq 0, \end{cases}$$

where v is the vertical velocity of one end of the shock absorber relative to the other end. A linear approximation over the range $-5 \leqq v \leqq 5$ is taken as

$$\hat{F}(k, v) = kv,$$

where k is chosen to minimize

$$G(k) = \int_{-5}^{5} \left[F(v) - \hat{F}(k, v)\right]^2 dv.$$

Determine the optimal value for k, subject to the constraints $0 \leqq k \leqq 50$.

2.19 Flight performance for aircraft may often be determined under the assumption that the acceleration forces of the aircraft are negligible (Miele, 1959). Consider motion in the vertical plane. If the thrust is aligned with the velocity vector, then the relationships between thrust T, drag D, weight W, flight path angle γ, and lift L are obtained by simple summation of forces:

$$T - D - W \sin\gamma = 0$$
$$L - W \cos\gamma = 0.$$

Lift and drag are obtained from

$$L=\tfrac{1}{2}\rho V^2 S C_L$$

$$D=\tfrac{1}{2}\rho V^2 S C_D$$

where ρ is the air density, V is the air speed, S is the wing area, C_L is the lift coefficient, and C_D is the drag coefficient. Lift and drag coefficients are related and for subsonic flight are often approximated by

$$C_D=C_{D0}+KC_L^2$$

where C_{D0} and K are constants. In this case it follows that

$$D=\tfrac{1}{2}\rho V^2 S C_{D0}+\frac{KW^2\cos^2\gamma}{\tfrac{1}{2}\rho V^2 S}$$

or

$$D=K_1V^2+\frac{K_2\cos^2\gamma}{V^2}$$

where $K_1=\tfrac{1}{2}\rho S C_{D0}$ and $K_2=2KW^2/\rho S$. If thrust T is a function of velocity only, show that to maximize the rate of climb ($V\sin\gamma$) for the aircraft at a given altitude (density) it is necessary that

$$\frac{\partial}{\partial V}[(T-D)V]=\frac{2K_2}{V^2}\sin^2\gamma.$$

2.20 If the flight path angle is small, show that the velocity for maximum rate of climb for the aircraft in Exercise 2.19 with constant thrust $T=T_0$ is given by

$$2V^2=\frac{T_0}{3K_1}+\left\{\left[\frac{T_0}{3K_1}\right]^2+\frac{4}{3}\frac{K_2}{K_1}\right\}^{1/2}.$$

2.21 A rocket engine is to be designed to propel a glider from the end of a runway at a constant speed V to an altitude of 1000 ft. Determine the mass flow rate of the engine β and the speed of the glider V that minimizes total fuel consumed. Assume $D=K_1V^2+K_2/V^2$, $T=\beta V_e$, $L=K_3\alpha V^2$, $\alpha_{\min}\leqq\alpha\leqq\alpha_{\max}$, where D is drag, T is thrust, L is lift, α is angle of attack, and K_1, K_2, K_3, V_e, $\alpha_{\min}$, and $\alpha_{\max}$ are all constants.

2.22 Determine the velocity at which an aircraft should fly in order to minimize the turning radius of a turn made in the horizontal plane. The aircraft cannot "pull" an acceleration of more than 4 times gravitational acceleration. Assume that the lift L is related to speed V and angle of attack α by $L=K_3\alpha V^2$ where K_3 is a constant and $\alpha\leqq\alpha_{\max}$.

Chapter Three

VECTOR MINIMIZATION

3.1 INTRODUCTION

When formulating practical optimization problems, particularly in such areas as "tradeoff" analysis, we often have to consider more than a single "cost" criterion associated with given choices for the decision variables $y \in Y$. When these cost criteria, say $G_1(y), \ldots, G_r(y)$, are noncommensurable, the optimization problem becomes one of minimizing a vector, that is

$$\min_{y \in Y} G(y) \tag{3.1}$$

subject to

$$g_i(y)=0 \qquad \forall i \in N=\{1,\ldots,n\} \tag{3.2}$$

$$h_j(y) \geqq 0 \qquad \forall j \in Q=\{1,\ldots,q\} \tag{3.3}$$

where $G(y)=[G_1(y),\ldots,G_r(y)]^T$ is now an r-dimensional vector instead of a scalar.

Given a multiple-criteria or vector minimization problem, what is meant by "minimizing a vector?" Attempting to define a vector minimal point as one at which all components of the cost vector are simultaneously minimized would not be an adequate generalization since such "utopia" points (Yu and Leitmann, 1976) are seldom attainable. For example, the point y that minimizes $G_1(y)$ need not minimize $G_2(y)$. Similarly, transforming the vector minimization problem into a scalar problem by minimizing some scalar-valued function of the vector cost components would simply ignore the basic question. However, scalarization can be a powerful technique for finding vector minima once a solution concept has been defined. The fundamental problem is to formulate a definition of a vector minimum when the components of the cost vector (e.g., apples and oranges) are noncommensurable. There are a number of rationales that can be used to approach this problem. A comprehensive survey of multicriteria optimization up to the year 1960 is given by Stadler (1979). The most widely used rationale studied by recent authors (e.g., DaCunha and Polak, 1967; Leitmann et al., 1972; Lin, 1976a, b, 1977; Yu, 1976; Vincent and Leitmann, 1970; Zadeh, 1963), is due to the engineer/economist Pareto (1896). This is the concept of an *undominated* or *Pareto-minimal* solution. Under this concept the cost vector $G(y^1)$ evaluated at $y^1 \in Y$ *dominates*, or is "better" than, the cost vector $G(y^2)$ evaluated at $y^2 \in Y$ if

$$G_i(y^1) \leqq G_i(y^2) \qquad \forall i \in R$$

and

$$G_j(y^1) < G_j(y^2) \quad \text{for at least one } j \in R$$

where $R = \{1, \ldots, r\}$.

It is clear that the point y^1 is preferable to y^2, since all cost components are no worse at y^2 and at least one cost component is smaller at y^1 than at y^2. Any choice of $y^* \in Y$ used to evaluate the cost vector should be such that y^* is not dominated by any other point $y \in Y$. Suppose it were possible to find the set of undominated or Pareto-minimal points. If $y \in Y$ is not a member of this set, then it is dominated by at least one point in the Pareto-minimal set.

Example 3.1 Let Y be the set of real numbers and (see Figure 3.1) let the cost vector $G(y)$ have the components

$$G_1(y) = y^2$$

$$G_2(y) = (y-2)^2.$$

Even though no "trade-off" relationship (e.g., "cost" $= aG_1 + bG_2$) is postulated between $G_1(\cdot)$ and $G_2(\cdot)$, the cost vector $G(y) = [G_1(y), G_2(y)]^T$ evaluated at a point such as $y = -1$ is clearly inferior to the cost vector evaluated at a point such as $y = 0$, since $G_i(0) < G_i(-1)$, $i = 1, 2$. In fact, at each point $y \in (-\infty, 0) \cup (2, +\infty)$ a (small) change δy in y exists such that $G_i(y + \delta y) < G_i(y)$ $\forall i = 1, 2$. Therefore, the points $y \in (-\infty, 0) \cup (2, +\infty)$ are not Pareto-minimal. Conversely, at every point $y \in (0, 2)$ any (small) change in y increases at least one component of the cost vector even though it decreases the other component. The points $y \in [0, 2]$ are the Pareto-minimal (undominated or vector minimal) points for this example.

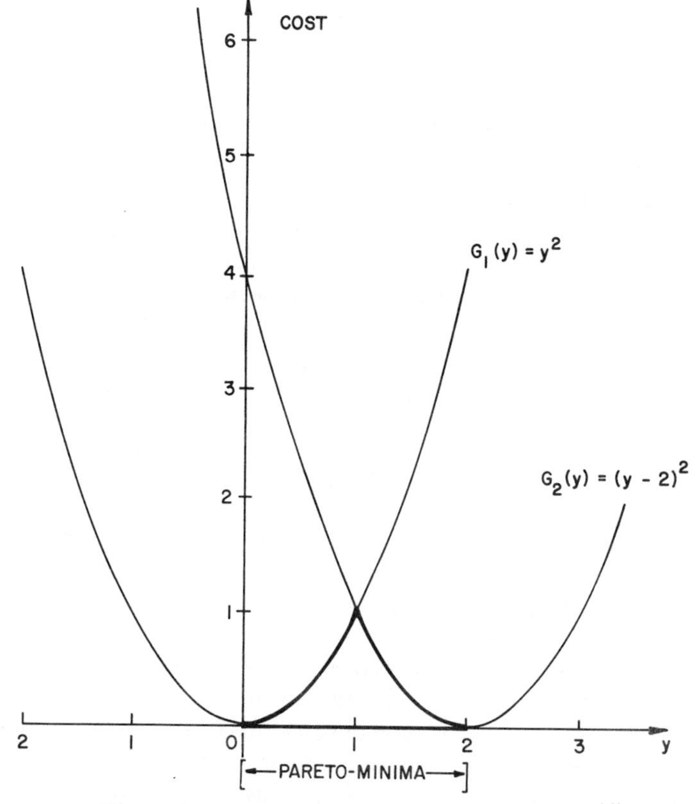

Figure 3.1. Pareto-minimal points for Example 3.1.

3.2 PARTIAL ORDER

The general question of preference for the cost vector evaluated at one point over the cost vector evaluated at another point implies the existence of some specified vector ordering relation. In seeking minima of a scalar function, the inequality relation $\leqq$ defines a complete order on the set of real numbers. For vector minimization, the ordering relation defined below is a strict partial order rather than a complete order.

For vector ordering we introduce the following notation: If G^1 and G^2 are two vectors of the same dimension with scalar components $[G_1^1, \ldots, G_r^1]^T$ and $[G_1^2, \ldots, G_r^2]^T$, respectively, then:

(i) $G^1 < G^2$ if and only if $G_i^1 < G_i^2 \; \forall i \in R$;

(ii) $G^1 \leqq G^2$ if and only if $G_i^1 \leqq G_i^2 \; \forall i \in R$;

(iii) $G^1 = G^2$ if and only if $G_i^1 = G_i^2 \ \forall i \in R$;

(iv) $G^1 \lessdot G^2$ if and only if $G_i^1 \leqq G_i^2 \ \forall i \in R$ and $G_j^1 < G_j^2$ for at least one $j \in R$.

In vector notation, Condition (iv) may be written as:

(iv) $G^1 \lessdot G^2$ if and only if $G^1 \leqq G^2$ and $G^1 \neq G^2$,

which defines a strict partial order on the set of cost vectors. The notation $\lessdot$ is due to Lin (1976a, b, 1977) and its use accents its principal difference with respect to the vector partial order $\leqq$; namely, $\lessdot$ (*partially less than*) implies strict inequality on at least one component.

3.3 PARETO-MINIMAL POINTS

DEFINITION 3.1 A point $y^* \in Y$ is a *Pareto-minimal point* for the vector-valued function $G(y) = [G_1(y), \ldots, G_r(y)]^T$ if and only if there does *not* exist a point $y \in Y$ such that $G(y) \lessdot G(y^*)$.

The Pareto concept for vector minimization is fundamentally different from the concept of a scalar minimum. In the case of a scalar minimum we seek a single point that minimizes the function. The concept of Pareto-minima is not to seek a single, "best" point, but instead to eliminate those points that are clearly inferior. Moreover, the vector minimization problem is a problem formulation based on less information than the scalar case. With a vector cost function the components are treated as being noncommensurable. In order to formulate a scalar problem from vector cost components, additional information concerning the decision-making problem is needed. Namely, some mapping of the cost vector into a cost scalar must be specified (e.g., weighted sum).

The Pareto concept of eliminating clearly inferior points based on noncommensurable cost components tends to generate continuous sets of local Pareto-minima rather than isolated minimal points. These sets, depending on the problem being addressed, may be of any dimension up to and including the dimension of the decision space itself. [For a discussion of existence of Pareto-minimal points in abstract spaces, see Cesari (1976)]. In numerous problems the set of Pareto-minimal points may be the entire decision space. For example, suppose that $G_1(y) = -G_2(y)$; then every admissible point $y \in Y$ would be a Pareto-minimal point.

Once the set of Pareto-minimal points has been found for a particular problem, additional information is required in order to select the "best" Pareto-minimal point. This is the subject of investigations in the general area of compromise solutions and bargaining [see Salukvadze (1979); Yu and Leitmann (1976)].

We now present a fundamental result for Pareto-minimal points in terms of global characterizations in the *cost space* defined by

$$\Xi \triangleq \{G(y) \in E^r \mid y \in Y\}.$$

Theorem 3.1 (Contact Theorem) A vector $G^* \in \Xi$ is Pareto-minimal if and only if

$$\Xi \cap C(G^*) = \{G^*\}$$

where

$$C(G^*) \triangleq \{G \in E^r \mid G = G^* + \mu,\ \mu \leqq 0\}$$

is the closed negative orthant centered at G^*.

Proof By definition $G^* \in \Xi$ is Pareto-minimal if and only if there does not exist a $G \in \Xi$ such that $G \leqslant G^*$. In other words, G^* is Pareto-minimal if and only if there does not exist a vector $\mu \leqslant 0$ such that

$$G \triangleq G^* + \mu \in \Xi.$$

Thus G^* is Pareto-minimal if and only if

$$\Xi \cap \{G \in E^r \mid G = G^* + \mu,\ \mu \leqslant 0\} = \varnothing$$

Taking the union of both sides of this relation with respect to G^* yields

$$\Xi \cap \{G \in E^r \mid G = G^* + \mu,\ \mu \leqq 0\} = \{G^*\}. \qquad \blacksquare$$

To illustrate the contact theorem consider the problem formulated in Example 3.1. The set of admissible cost vectors

$$\Xi = \left\{G \in E^2 \mid G = \left[y^2, (y-2)^2\right]^T, y \in E^1\right\}$$

is shown in Figure 3.2. The Pareto-minimal solutions are shown in bold and are determined by direct application of the contact theorem.

Practical utilization of the contact theorem is limited by a requirement to implicitly or explicitly plot the set Ξ of admissible cost vectors. The theorem, however, does provide a clear geometric illustration of the concept

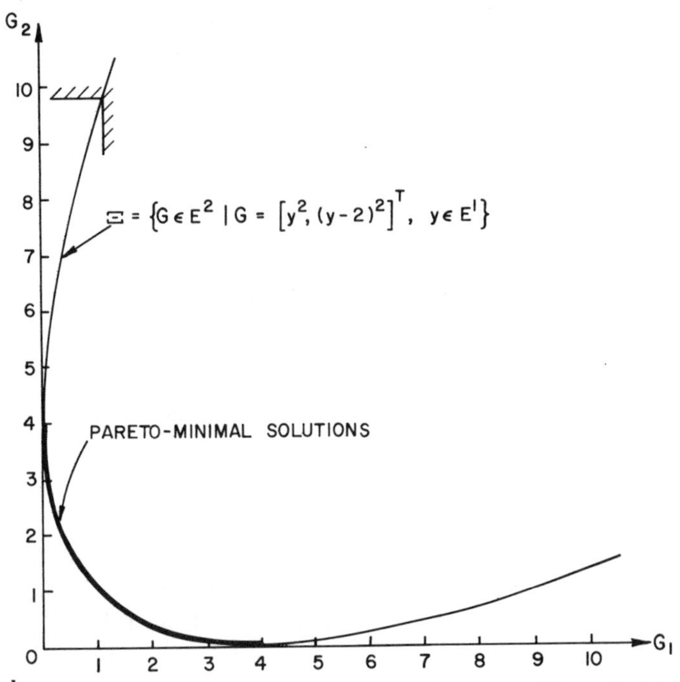

Figure 3.2. Application of the contact theorem to Example 3.1.

of a Pareto-minimal point. Fortunately, determination of the entire set Ξ of admissible cost vectors is generally not required. Additional global properties of Pareto-minimal points are discussed in Section 3.6.

3.4 LOCAL VECTOR MINIMA

DEFINITION 3.2 A point $y^* \in Y$ is a *local Pareto-minimal* point for the vector-valued function $G(y)$ if and only if there exists a ball B centered at y^* such that there does *not* exist a point $y \in B \cap Y$ with $G(y) \leqslant G(y^*)$.

The definition of a Pareto-minimal point differs in form from its scalar counterpart [i.e., $y^* \in Y$ is a minimal point for the scalar function $f(\cdot)$ if and only if $f(y^*) \leqq f(y)\ \forall y \in Y$]. The difference is made clear by examining the *utopia point* condition $G(y^*) \leqq G(y)\ \forall y \in Y$ for a vector function $G(\cdot)$. If it exists, the utopia point is indeed a Pareto-minimal point, but such a restrictive definition eliminates those "trade-off" or noncomparable points [where for a change in y some components of $G(y)$ decrease while others increase] that are allowed under the definition of Pareto-minimal points.

Given a cost vector $G=[G_1,\dots,G_r]^T$, let $\{y^*\}_i$ denote the set of points at which component G_i takes on a *proper* minimum. Then each point in the set $\cup_{i=1}^{r}\{y^*\}_i$ is a Pareto-minimal point for the cost vector G. In other words, any point that is a *proper* minimum for a component of G is a Pareto-minimal point (see Lemma 3.3). In general, however, these are by no means all of the Pareto-minimal points for G.

From the definition of a local Pareto-minimal point y^* there exists a ball B centered at y^* such that

$$G(y) \leqslant G(y^*) \tag{3.4}$$

does *not* have a solution $y \in B \cap Y$; that is, for all $y \in B \cap Y$ either:

$$G(y) = G(y^*) \tag{3.5}$$

or

$$G_i(y) > G_i(y^*) \tag{3.6}$$

for at least one $i \in \{1,\dots,r\}$.

This latter observation could be used as an alternate definition of local Pareto-minimal points.

Local conditions are inherently related to the structure of the *decision space* Y. As in Chapters 1 and 2, we consider the case where Y is defined by

$$g(y) = 0 \tag{3.7}$$

$$h(y) \geqq 0 \tag{3.8}$$

where $g(\cdot): E^m \to E^n$ and $h(\cdot): E^m \to E^q$ are C^1.

We now present a basic necessary condition for local Pareto-minimality in terms of the tangent cone to Y at a point y^*.

Lemma 3.1 Let $y^* \in Y$ be a local Pareto-minimal point for a vector-valued cost function $G(\cdot): Y \to E^r$, which is C^1 at y^*. Then it is necessary that there does *not* exist a vector $e \in T$ tangent to Y at y^* such that

$$\frac{\partial G(y^*)}{\partial y} e < 0. \tag{3.9}$$

Proof Suppose there exists an $e \in T$ such that the vector-valued condition (3.9) is satisfied and let $\delta y(\cdot)$ be any function that generates the tangent vector e. Then, for any scalar $\alpha>0$, $y^*+\alpha\,\delta y(\alpha)\to e$ as $\alpha\to 0$. From the first-order approximation theorem we have

$$G[y^*+\alpha\,\delta y(\alpha)]=G(y^*)+\frac{\partial G(y^*)}{\partial y}\alpha\,\delta y(\alpha)+R(\alpha) \tag{3.10}$$

where

$$\frac{\partial G}{\partial y}=\begin{bmatrix} \dfrac{\partial G_1}{\partial y_1} & \cdots & \dfrac{\partial G_1}{\partial y_m} \\ \vdots & \ddots & \vdots \\ \dfrac{\partial G_r}{\partial y_1} & \cdots & \dfrac{\partial G_r}{\partial y_m} \end{bmatrix}$$

and $R(\cdot)$ is a vector-valued function such that $R(\alpha)/\alpha\to 0$ as $\alpha\to 0$. Dividing by $\alpha>0$ and taking the limit as $\alpha\to 0$, we have

$$\lim_{\alpha\to 0}\frac{1}{\alpha}\{G[y^*+\alpha\,\delta y(\alpha)]-G(y^*)\}=\frac{\partial G(y^*)}{\partial y}e. \tag{3.11}$$

But (3.9) and (3.11) imply that

$$G[y^*+\alpha\,\delta y(\alpha)]<G(y^*) \tag{3.12}$$

for sufficiently small $\alpha>0$, which contradicts the Pareto-minimality of y^*. ■

Note that the vector inequality in (3.9) cannot be replaced by $\leqslant$ and still guarantee that (3.12) would hold for a sufficiently small $\alpha>0$. For example, suppose (3.9) were to read

$$\frac{\partial G(y^*)}{\partial y}e\leqslant 0.$$

Then in the proof of Lemma 3.1 some $i\in R$ could exist such that

$$\frac{\partial G_i(y^*)}{\partial y}=0$$

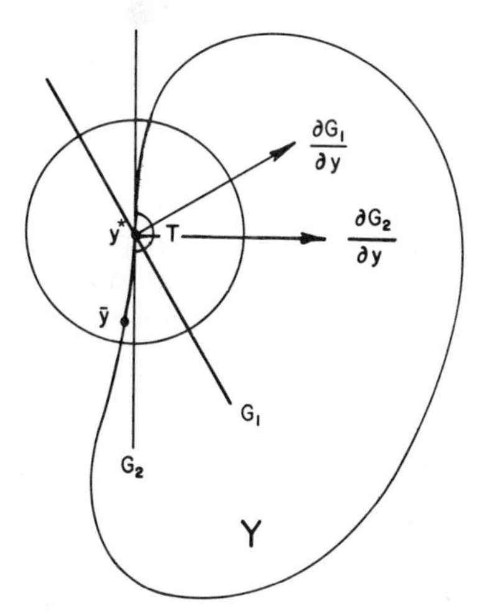

Figure 3.3. A point y^* that satisfies Lemma 3.1 but is not Pareto-minimal.

and the second-order approximation theorem would yield

$$\lim_{\alpha\to 0}\frac{1}{\alpha}\{G_i[y^*+\alpha\delta y(\alpha)]-G_i(y^*)\}=\tfrac{1}{2}e^T\frac{\partial^2 G_i(y^*)}{\partial y^2}e$$

from which no contradiction need follow. Thus Lemma 3.1 is only a first-order approximation to the contact theorem.

It should also be noted that even though Lemma 3.1 is a powerful necessary condition, it is not a necessary and sufficient condition, whereas the contact theorem is. Figure 3.3 illustrates a situation in which a point satisfies Lemma 3.1 but would not satisfy the contact theorem, even locally. Regardless of how small the ball B is chosen in Figure 3.3, there always exists a point $\bar{y}\in B\cap Y$ such that $G(\bar{y})\leqslant G(y^*)$.

As with scalar optimization problems, sufficiency conditions can be employed to determine local Pareto-minimal points. The following two lemmas are sufficiency conditions applicable to a C^1 cost vector.

Lemma 3.2 A point $y^*\in Y$ is a local Pareto-minimal point for a vector-valued cost function $G(\cdot): Y\to E^r$ that is C^1 at y^* if for each vector $e\in T$ there exist two components of the vector

$$\beta=\frac{\partial G(y^*)}{\partial y}e$$

that are of opposite sign.

Proof From the first-order approximation theorem

$$\lim_{\alpha\to 0}\frac{1}{\alpha}\{G[y^*+\alpha\,\delta y(\alpha)]-G(y^*)\}=\frac{\partial G(y^*)}{\partial y}e=\beta$$

where $\delta y(\cdot)$ is any generator of e. From the opposite signs hypothesis $\beta \nleq 0$ holds for each $e\in T$. Thus there exists no $y=y^*+\alpha\,\delta y(\alpha)$ in $B\cap Y$ such that $G(y)\leqslant G(y^*)$, which implies that y^* is a local Pareto-minimal point. ■

Lemma 3.3 A point $y^*\in Y$ is a local Pareto-minimal point for a vector-valued function $G(\cdot): Y\to E$ that is C^1 at y^* if for some $i\in\{1,\dots,r\}$

$$\frac{\partial G_i(y^*)}{\partial y}e>0 \tag{3.13}$$

for all nonzero $e\in T$ where T is the tangent cone to Y at y^*.

Proof Let $\delta y(\cdot)$ generate a nonzero tangent vector $e\in T$. From (3.11) and (3.13)

$$G_i[y^*+\alpha\,\delta y(\alpha)]>G_i(y^*)$$

for all sufficiently small $\alpha>0$. Therefore, for some sufficiently small ball B about y^*, there does not exist a $y\in B\cap Y$ such that $G(y)\leqslant G(y^*)$. Thus y^* is a local Pareto-minimal point. ■

From Lemma 3.3 we see that any point at which a component of the cost vector takes on a proper local minimum is a local Pareto-minimal point.

Example 3.2 Consider $y\in E^1$ and let $G(\cdot)=[G_1(\cdot), G_2(\cdot)]^T$ be given by

$$G_1(\cdot)=y^2$$

$$G_2(\cdot)=(y-2)^2.$$

Then $T=E^1$. Applying Lemma 3.2,

$$\beta=\frac{\partial G(y^*)}{\partial y}e=\begin{bmatrix}2y^*e\\ 2(y^*-2)e\end{bmatrix}=\begin{bmatrix}\beta_1\\ \beta_2\end{bmatrix}.$$

For the components of β to have opposite signs we have

$$\beta_1\beta_2 = 4e^2 y^*(y^* - 2) < 0,$$

which implies $y^* \in (0, 2)$.

As a consequence of (2.10), (2.11), and Lemma 3.1, we have the following generalization of the necessary condition of Schmitendorf (1972) at a regular local Pareto-minimal point.

Lemma 3.4 Let y^* be a regular local Pareto-minimal point for a vector-valued cost function $G(\cdot): Y \to E^r$ that is C^1 at y^*. Then it is necessary that

$$\operatorname{rank}\begin{bmatrix} \dfrac{\partial G}{\partial y} \\ \dfrac{\partial g}{\partial y} \\ \dfrac{\partial \hat{h}}{\partial y} \end{bmatrix}_{y^*} < \operatorname{rank}\begin{bmatrix} \dfrac{\partial G}{\partial y} & -\alpha \\ \dfrac{\partial g}{\partial y} & 0 \\ \dfrac{\partial \hat{h}}{\partial y} & \beta \end{bmatrix}_{y^*}$$

$\forall \alpha > 0$, $\beta \geqq 0$, where $\alpha \in E^r$, $\beta \in E^{q*}$, and $\hat{h}(\cdot) = [h_1(\cdot), \ldots, h_{q*}(\cdot)]^T$ is the vector of active inequality constraints at y^*.

Proof Consider the system of equations

$$Ae = b$$

where

$$A = \begin{bmatrix} \dfrac{\partial G}{\partial y} \\ \dfrac{\partial g}{\partial y} \\ \dfrac{\partial \hat{h}}{\partial y} \end{bmatrix}_{y^*}$$

$$b = \begin{bmatrix} -\alpha \\ 0 \\ \beta \end{bmatrix}$$

for $\alpha > 0$, $\beta \geqq 0$ where $\alpha \in E^r$, $\beta \in E^{q^*}$. If $Ae = b$ has a solution e, then from (2.10) and (2.11), $e \in T$ and

$$\frac{\partial G(y^*)}{\partial y} e = -\alpha < 0,$$

which would violate Lemma 3.1. Therefore, $Ae = b$ has no solution at a regular local Pareto-minimal point y^*. Thus from linear algebra rank $[A] \neq \text{rank}[A \,\vdots\, b]$, which implies rank $[A] < \text{rank}[A \,\vdots\, b]$. ■

Example 3.3 Suppose $y \in E^2$ with two cost components

$$G_1(\cdot) = -y_1 y_2$$

$$G_2(\cdot) = y_1^2 + y_2$$

and a single equality constraint

$$g(\cdot) = y_1 + y_2 = 0.$$

Then Lemma 3.4 requires

$$\text{rank}[A] < \text{rank}[A \,\vdots\, b]$$

for all $\alpha_1 > 0$, $\alpha_2 > 0$ where

$$A = \begin{bmatrix} \dfrac{\partial G}{\partial y} \\ \dfrac{\partial g}{\partial y} \end{bmatrix} = \begin{bmatrix} -y_2 & -y_1 \\ 2y_1 & 1 \\ 1 & 1 \end{bmatrix}$$

and

$$b = \begin{bmatrix} -\alpha_1 \\ -\alpha_2 \\ 0 \end{bmatrix}.$$

Since elementary row (or column) operations do not change the rank of a matrix, the rank of $[A \,\vdots\, b]$ and the rank of A may be determined by using elementary row operations to transform $[A \,\vdots\, b]$ to an upper triangular matrix. The rank of the resulting matrix is then equal to the number of nonzero rows. We obtain

$$[A \mid b] = \begin{bmatrix} -y_2 & -y_1 & -\alpha_1 \\ 2y_1 & 1 & -\alpha_2 \\ 1 & 1 & 0 \end{bmatrix}$$

$$\to \begin{bmatrix} -y_2 & -y_1 & -\alpha_1 \\ 0 & 1-2y_1 & -\alpha_2 \\ 0 & y_2-y_1 & -\alpha_1 \end{bmatrix}$$

$$\to \begin{bmatrix} y_1 & -y_1 & -\alpha_1 \\ 0 & 1-2y_1 & -\alpha_2 \\ 0 & 0 & -\alpha_1+2(\alpha_1-\alpha_2)y_1 \end{bmatrix}$$

where we have used the equality constraint in the last step.

Suppose $y_1(1-2y_1) \neq 0$. Then rank $[A]=2$ and Lemma 3.4 requires that

$$-\alpha_1+2(\alpha_1-\alpha_2)y_1 \neq 0$$

hold for all $\alpha_1>0$, $\alpha_2>0$. Thus dividing by $\alpha_1>0$,

$$-1+2(1-\beta)y_1 \neq 0$$

must hold for all $\beta>0$. Therefore, since $\beta=1$ is a valid choice, we require

$$-1+2(1-\beta)y_1<0$$

$\forall \beta>0$. Hence

$$y_1<\frac{1}{2(1-\beta)} \qquad \forall \beta \in (0,1)$$

$$y_1>\frac{1}{2(1-\beta)} \qquad \forall \beta>1.$$

Therefore,

$$0<y_1<\tfrac{1}{2}.$$

We return now to the case where $y_1(1-2y_1)=0$. For $y_1=0$, or $y_1=\frac{1}{2}$, rank $[A]=1<\text{rank}[A\,|\,b]=2$ since $\alpha_1>0$ and $\alpha_2>0$. Thus the (candidate) Pareto-minimal points are given by $0\leqq y_1\leqq\frac{1}{2}$, $y_2=-y_1$.

We now turn to a discussion of the geometry of the tangent cone at a local Pareto-minimal point y^* and its relation to the cone generated by the gradients $\partial G_i(y^*)/\partial y$, $i=1,\ldots,r$.

At any point $y^*\in Y$, with $G(\cdot)$, $g(\cdot)$, and $h(\cdot)\,C^1$ at y^*, we define the following convex cones:

$$C=\left\{y\in E^m\,|\,y^T=\eta^T\frac{\partial G(y^*)}{\partial y},\ \eta\geqslant 0\right\} \tag{3.14}$$

$$\mathring{C}_D=\left\{e\in E^m\,|\,\frac{\partial G(y^*)}{\partial y}e<0\right\} \tag{3.15}$$

$$(\mathring{C}_D)^*=\left\{y\in E^m\,|\,y^T=-\eta^T\frac{\partial G(y^*)}{\partial y},\ \eta\geqslant 0\right\} \tag{3.16}$$

$$K=\left\{y\in E^m\,|\,y^T=\lambda^T\frac{\partial g(y^*)}{\partial y}+\mu^T\frac{\partial h(y^*)}{\partial y},\ \mu\geqq 0 \text{ with } \mu^T h(y^*)=0\right\} \tag{3.17}$$

$$K^*=\left\{e\in E^m\,|\,y^Te\geqq 0\ \forall y\in K\right\}. \tag{3.18}$$

The cone C is called the *cost cone* and consists of all nonzero nonnegative linear combinations of the gradient vectors $\{\partial G_i(y^*)/\partial y\}$, $i=1,\ldots,r$. $\mathring{C}_D$ is the open dual of C and $(\mathring{C}_D)^*$ is the polar to $\mathring{C}_D$. We have taken the liberty of calling C, $\mathring{C}_D$, and $(\mathring{C}_D)^*$ cones even though they do not generally contain the zero vector.

The cone K is the *constraint cone* (see Chapter 2) and K^* is the polar to K. The cone K consists of all nonnegative linear combinations of the constraint gradients $\{\partial g_i/\partial y, -\partial g_i/\partial y, \partial h_j/\partial y\}_{y^*}$, $i\in N$, $j\in Q^*=\{1,\ldots,q^*\}$. In the unconstrained case ($N=Q=Q^*=\varnothing$) we take $K=\{0\}$ and $K^*=E^m$.

Figure 3.4 is an illustration of the cost and constraint cones at a point y for the case of two cost components and two inequality constraints. From Lemma 2.5 $K^*=T$ at a regular point. We now show that a point such as y in Figure 3.4 is not a Pareto-minimal point.

In terms of the open dual to the cost cone, we have the following fundamental relationship between the cost cone and the tangent cone at a local Pareto-minimal point.

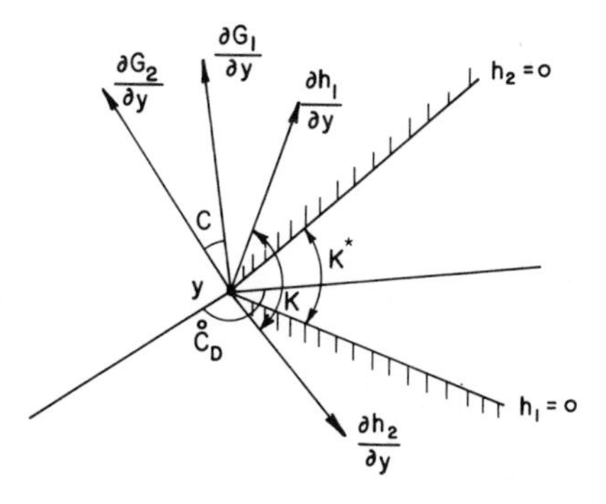

Figure 3.4. Cost and constraint cones at a non-Pareto-minimal point.

Lemma 3.5 If $y^* \in Y$ is a local Pareto-minimal point for a vector-valued function $G(\cdot): Y \to E^r$ that is C^1 at y^*, then

$$\mathring{C}_D \cap T = \varnothing \tag{3.19}$$

where T is the tangent cone to Y at y^* and $\mathring{C}_D$ is defined by (3.15).

Proof The lemma follows immediately from Lemma 3.1 and the definition of $\mathring{C}_D$. ■

By applying Lemma 3.5 to the situation illustrated in Figure 3.4, we see that since $K^* = T$ (Lemma 2.5) the point y is not a Pareto-minimal point because $\mathring{C}_D \cap T \neq \varnothing$ at y. Figure 3.5 illustrates a situation that satisfies Lemma 3.5. Note that Lemma 3.5 applies to an arbitrary tangent set T. In particular, Lemma 3.5 is not restricted to regular points, nor is it restricted to constraint sets Y of the form (3.7)–(3.8).

We now develop a modification of Farkas' lemma applicable to the open dual cone $\mathring{C}_D$.

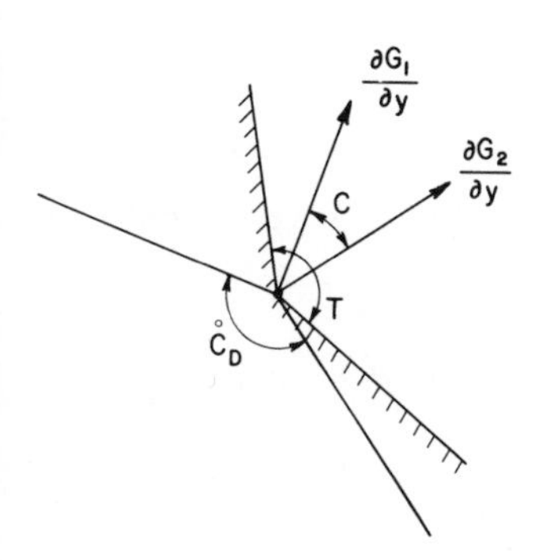

Figure 3.5. An illustration of the geometry of Lemma 3.5.

Lemma 3.6 Let

$$\mathring{C}_D = \left\{ e \in E^m \mid \frac{\partial G(y^*)}{\partial y} e < 0 \right\} \tag{3.20}$$

and

$$(\mathring{C}_D)^* = \left\{ y \in E^m \mid y^T = -\eta^T \frac{\partial G(y^*)}{\partial y}, \; \eta \geqslant 0 \right\}. \tag{3.21}$$

Then $e \in \mathring{C}_D$ if and only if $y^T e > 0 \; \forall y \in (\mathring{C}_D)^*$.

Proof If $e \in \mathring{C}_D$, then by definition

$$\frac{\partial G(y^*)}{\partial y} e < 0.$$

Therefore,

$$\left[-\eta^T \frac{\partial G(y^*)}{\partial y} \right] e > 0 \qquad \forall \eta \geqslant 0,$$

which implies $y^T e > 0 \; \forall y \in (\mathring{C}_D)^*$.

Conversely, if $y^T e > 0 \; \forall y \in (\mathring{C}_D)^*$, then from the definition of $(\mathring{C}_D)^*$,

$$-y^T e = \eta^T \left[\frac{\partial G(y^*)}{\partial y} e \right] < 0 \qquad \forall \eta \geqslant 0,$$

which implies

$$\frac{\partial G(y^*)}{\partial y} e < 0.$$

Therefore, $e \in \mathring{C}_D$. ∎

From Lemma 3.6 and Farkas' lemma (see Chapter 1) we establish the following theorem.

Theorem 3.2 (Theorem of the Alternative) Let C, $\mathring{C}_D$, $(\mathring{C}_D)^*$, K, and K^* be defined by (3.14)–(3.18). Then

$$\mathring{C}_D \cap K^* = \varnothing \Leftrightarrow C \cap K \neq \varnothing. \tag{3.22}$$

Proof Suppose $K^* = \varnothing$. Then $K = (K^*)^* = E^m$, which implies $C \cap K = C \neq \varnothing$ since $C \neq \varnothing$. Now, consider the case where $K^* \neq \varnothing$ and let e be an arbitrary vector in K^*. From the hypothesis $\mathring{C}_D \cap K^* = \varnothing$ we have $e \notin \mathring{C}_D$. Therefore, from Lemma 3.6, for each $e \in K^*$ there exists a vector $y_e \in (\mathring{C}_D)^*$ such that

$$y_e^T e \leqq 0. \tag{3.23}$$

From the definitions of $(\mathring{C}_D)^*$ and C, $y_e \in (\mathring{C}_D)^*$ if and only if $-y_e \in C$. We now show that $-y_e \in K$, thereby establishing that $C \cap K \neq \varnothing$.

Suppose $-y_e \notin K$. Then from Farkas' lemma there exists a vector $e \in K^*$ such that $(-y_e)^T e < 0$, that is,

$$y_e^T e > 0. \tag{3.24}$$

But given $e \in K^*$, y_e is chosen to satisfy (3.23). Therefore, (3.23) and (3.24) are contradictory and we conclude that $-y_e \in C \cap K \neq \varnothing$.

Conversely, for $C \cap K \neq \varnothing$, choose $\bar{y} \in C \cap K$. From Farkas' lemma $\bar{y} \in K$ if and only if

$$\bar{y}^T e \geqq 0 \qquad \forall e \in K^*. \tag{3.25}$$

Now, by way of contradiction suppose $\mathring{C}_D \cap K^* \neq \varnothing$. Then there exists a vector $e \in \mathring{C}_D \cap K^*$. From Lemma 3.6, $e \in \mathring{C}_D$ if and only if $y^T e > 0$ $\forall y \in (\mathring{C}_D)^*$. But from (3.14) and (3.16) $y \in (\mathring{C}_D)^*$ if and only if $-y \in C$. Therefore, $e \in \mathring{C}_D$ if and only if

$$y^T e < 0 \qquad \forall y \in C. \tag{3.26}$$

In particular, if $e \in \mathring{C}_D \cap K^*$, then (3.26) must hold for $y = \bar{y} \in C \cap K$. That is, $\bar{y}^T e < 0$, which contradicts (3.25). Thus we have that $C \cap K \neq \varnothing \Rightarrow \mathring{C}_D \cap K^* = \varnothing$. ■

The following lemma provides a basis for determining local Pareto-minimal points.

Lemma 3.7 If $y^* \in Y$ is a regular local Pareto-minimal point for a vector-valued cost function $G(\cdot): Y \to E^r$ that is C^1 at y^*, then there exists a vector $\eta \gg 0$ such that

$$\eta^T \frac{\partial G(y^*)}{\partial y} e \geqq 0 \qquad \forall e \in T. \tag{3.27}$$

Proof From Lemma 2.5, $T=K^*$ at a regular point y^*. From Lemma 3.5 $\mathring{C}_D \cap T = \varnothing$ at a local Pareto-minimal point. Thus we have

$$\mathring{C}_D \cap K^* = \varnothing$$

at a regular local Pareto-minimal point y^*. It then follows from Theorem 3.2 that

$$C \cap K \neq \varnothing$$

at y^*. Thus there exists a vector $y \in K$ of the form

$$y = \eta^T \frac{\partial G(y^*)}{\partial y}, \qquad \eta \geqslant 0.$$

Now, from Farkas' lemma $y \in K$ if and only if $y^T e \geqq 0 \ \forall e \in K^*$. Since $K^* = T$, we have

$$\eta^T \frac{\partial G(y^*)}{\partial y} e \geqq 0$$

$\forall e \in T$. ■

Note that (3.27) would not hold for a tangent set such as the case illustrated in Figure 2.8*b*. From Lemma 2.4 the restriction that y^* be a regular point, with tangent vectors defined by (2.10) and (2.11), implies that T is a convex cone. For regular points then there exists a hyperplane normal to some vector $\beta \in E^m$ such that

$$\beta^T e \geqq 0 \qquad \forall e \in T. \tag{3.28}$$

If y^*, in addition to being a regular point, is also a Pareto-minimal point, then according to Lemma 3.7, the vector β in (3.28) may be chosen so that

$$\beta \in C$$

where C is the cost cone defined by (3.14).

From Lemma 2.1, condition (3.27) is the same as the necessary condition for $y^* \in Y$ to be a regular local minimal point for the scalar function $\eta^T G(y)$ for some appropriate $\eta \geqslant 0$. As we show in Section 3.7, it does not follow that Pareto-minimal points necessarily minimize $\eta^T G(y)$.

From Lemmas 3.5 and 2.5 and Theorem 3.2 we have the following necessary condition for regular local Pareto-minimal points.

Theorem 3.3 If $y^* \in Y$ is a regular local Pareto-minimal point for a vector-valued function $G(\cdot): Y \to E^r$ that is C^1 at y^*, then there exist vectors $\eta \in E^r$, $\lambda \in E^n$, $\mu \in E^q$ with $\eta \geqslant 0$, $\mu \geqq 0$ such that

$$\frac{\partial L(y^*, \eta, \lambda, \mu)}{\partial y} = 0 \tag{3.29}$$

$$g(y^*) = 0 \tag{3.30}$$

$$h(y^*) \geqq 0 \tag{3.31}$$

and

$$\mu^T h(y^*) = 0 \tag{3.32}$$

where

$$L(y, \eta, \lambda, \mu) = \eta^T G(y) - \lambda^T g(y) - \mu^T h(y). \tag{3.33}$$

Proof From Lemma 3.5 $\mathring{C}_D \cap T = \varnothing$. Thus from Lemma 2.5 $\mathring{C}_D \cap K^* = \varnothing$. The theorem then follows directly from (3.14), (3.17), and Theorem 3.2. ■

Note that (3.29) implies that

$$\operatorname{rank}\begin{bmatrix} \eta^T \dfrac{\partial G}{\partial y} \\ \dfrac{\partial g}{\partial y} \\ \dfrac{\partial \hat{h}}{\partial y} \end{bmatrix}_{y^*} = \operatorname{rank}\begin{bmatrix} \dfrac{\partial g}{\partial y} \\ \dfrac{\partial \hat{h}}{\partial y} \end{bmatrix}_{y^*} \tag{3.34}$$

for some $\eta \geqslant 0$ where $\hat{h}(y^*)$ is the vector of active inequality constraints at y^*. A function $h_j(y^*)$ is a component of $\hat{h}(y^*)$ if and only if $h_j(y^*) = 0$. In the absence of constraints, we have the following corollary.

Corollary 3.1 If y^* is a local Pareto-minimal point for a C^1 vector-valued function $G(\cdot): Y \to E^r$ in the absence of equality and inequality constraints, then the cost gradients $\{\partial G_i(y^*)/\partial y\}$ must be linearly dependent at y^*.

Example 3.4 Let $Y=E^2$ (no equality or inequality constraints) and consider a cost vector with two components

$$G_1(\cdot)=\tfrac{1}{2}y_1^2+y_2^2$$

$$G_2(\cdot)=\tfrac{1}{2}(y_1-1)^2+\tfrac{1}{2}(y_2-1)^2.$$

Then

$$\begin{aligned} L&=\eta^T G \\ &=\eta_1\left[\tfrac{1}{2}y_1^2+y_2^2\right]+\eta_2\left[\tfrac{1}{2}(y_1-1)^2+\tfrac{1}{2}(y_2-1)^2\right] \end{aligned}$$

and we have

$$\frac{\partial L}{\partial y_1}=0=\eta_1 y_1^*+\eta_2(y_1^*-1)$$

$$\frac{\partial L}{\partial y_2}=0=2\eta_1 y_2^*+\eta_2(y_2^*-1).$$

These relations may be written as

$$\begin{bmatrix} y_1^* & y_1^*-1 \\ 2y_2^* & y_2^*-1 \end{bmatrix}\begin{bmatrix} \eta_1 \\ \eta_2 \end{bmatrix}=\begin{bmatrix} 0 \\ 0 \end{bmatrix}.$$

In order to have a nonzero solution $\eta \geqslant 0$ we must have

$$y_1^*(y_2^*-1)-2y_2^*(y_1^*-1)=0$$

from which

$$y_1^*=\frac{2y_2^*}{1+y_2^*}.$$

From $\partial L/\partial y_2=0$ we have

$$y_2^*=\frac{\eta_2}{2\eta_1+\eta_2}$$

where $\eta=[\eta_1,\eta_2]^T \geqslant 0$. Thus

$$0 \leqq y_2^* \leqq 1.$$

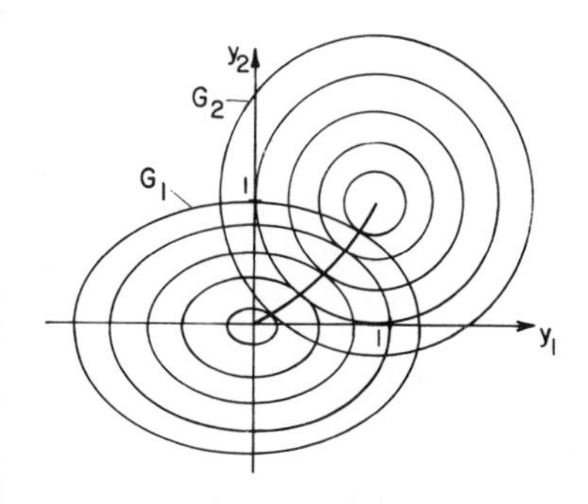

Figure 3.6 The Pareto-minimal set for Example 3.4.

The Pareto-minimal points (y_1^*, y_2^*) and some isocost curves for $G_1(\cdot)$ and $G_2(\cdot)$ are shown in Figure 3.6. Note that, in the absence of constraints, the Pareto-minimal points (y_1^*, y_2^*) are the points of tangency of the isocost curves of $G_1(\cdot)$ and $G_2(\cdot)$ between $(0,0)$, the minimal point for $G_1(\cdot)$, and $(1,1)$, the minimal point for $G_2(\cdot)$.

3.5 COMPROMISE SOLUTIONS

In this section we examine an alternate solution concept for multiple criteria problems, known as compromise solutions (Salukvadze, 1979: Yu and Leitmann, 1976).

DEFINITION 3.3 A point $G^\circ \in E^r$ is a *utopia point* if and only if for each $i=1,\ldots,r$

$$G_i^\circ = \inf\{G_i(y) | y \in Y\}.$$

Note that a utopia point G° is a single, unique point in the cost space. If G° is feasible, that is, if $G^\circ \in \Xi = \{G(y) \in E^r | y \in Y\}$, then G° corresponds to a point $y \in Y$ at which all components of $G(y)$ are simultaneously minimized. In general, however, G° is not feasible, as illustrated in Figure 3.7.

For the general case where the utopia point is not feasible, one reasonable approach to multicriteria optimization might be to seek a solution that is as "close" as possible to the utopia point. This idea forms the basis for the concept of compromise solutions.

Since the cost components are considered noncommensurable (e.g., different units) there is no *a priori* reason for restricting the discussion of "closeness" to the special case of the Euclidian norm, our usual measure of

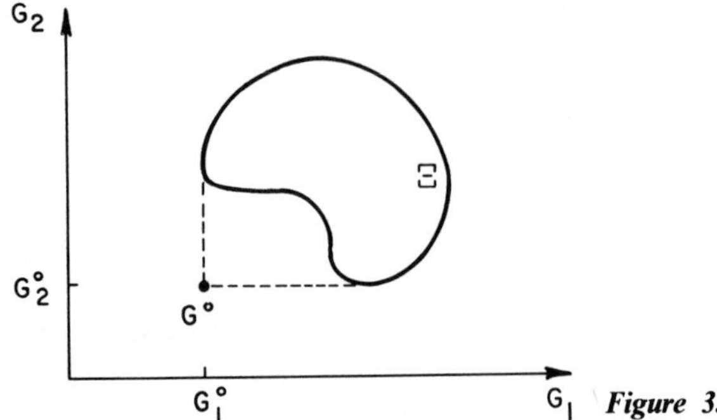

Figure 3.7. Utopia point.

distance. Thus we consider the following family of *regret functions*:

$$R_p(y)=\|G(y)-G^\circ\|_p=\left\{\sum_{i=1}^{r}\left[G_i(y)-G_i^\circ\right]^p\right\}^{1/p}$$

where $p \geqq 1$. For $p=2$ the regret function reduces to the Euclidian norm.

Example 3.5 Determine all compromise solutions for $p \geqq 1$ where

$$G_1(y)=\frac{1}{y}$$

$$G_2(y)=y^2$$

with scalar $y>0$. Since

$$\inf_{y>0}\frac{1}{y}=0$$

$$\inf_{y>0}\left[y^2\right]=0,$$

we have

$$G^\circ=[0,0]^T.$$

Defining

$$R_p(y)=\left\{\left[\frac{1}{y}-0\right]^p+\left[y^2-0\right]^p\right\}^{1/p}$$

$$=\left\{y^{-p}+y^{2p}\right\}^{1/p},$$

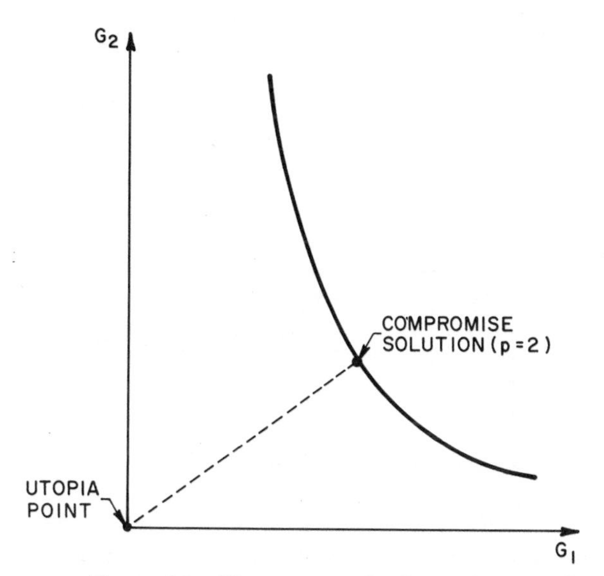

Figure 3.8. The geometry for Example 3.5.

we have the following necessary condition at a local minimal point for $R_p(\cdot)$:

$$\frac{\partial R_p}{\partial y}=0=\frac{1}{p}\left[-py^{-p-1}+2py^{2p-1}\right]\left[y^{-p}+y^{2p}\right]^{(1/p)-1}.$$

Therefore,

$$y=(2)^{-1/3p}.$$

The compromise solution with $p=2$ is illustrated in Figure 3.8.

3.6 GLOBAL VECTOR MINIMA

In this section we present some results associated with global Pareto-minimal points. We are concerned here with global characterizations in the cost space

$$\Xi \triangleq \{G(y)\in E^r \mid y\in Y\}$$

rather than in the decision space Y. The developments in this section parallel Lin (1976a, b, 1977).

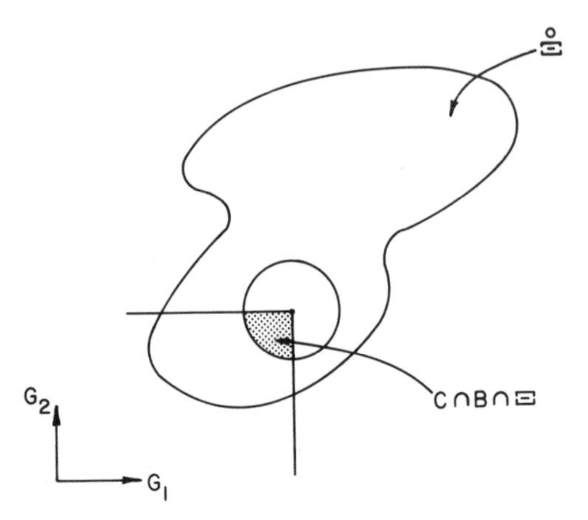

Figure 3.9. An illustration of the geometry for Theorem 3.4.

The contact theorem of Section 3.3 is a global result. The following theorem demonstrates that determination of the entire set Ξ of admissible cost vectors is not generally required in order to utilize the contact theorem.

Theorem 3.4 If $G^* \in \Xi$ is a Pareto-minimal point, then $G^* \in \partial \Xi$.

Proof Suppose $G^* \in \mathring{\Xi}$ (the interior of Ξ). Then (see Figure 3.9) there exists a ball B about G^* such that $B \cap \Xi \subset \mathring{\Xi}$. Hence $C(G^*) \cap [B \cap \Xi] - \{G^*\} \neq \varnothing$, which contradicts the contact theorem. ■

The following theorem (Schmitendorf, 1972) illustrates that Pareto-minimal points for a vector minimization problem may be found by solving a system of scalar minimization problems.

Theorem 3.5 A point $G^* \in \Xi$ is Pareto-minimal if and only if for each $j \in \{1, \ldots, r\}$

$$G_j^* \leqq G_j \qquad \forall G \in \Xi_j$$

where

$$\Xi_j = \{G \in \Xi \mid G_i \leqq G_i^* \; \forall i = 1, \ldots, r; \; i \neq j\}.$$

Proof Suppose G^* is Pareto-minimal with $G_j^* > G_j$ for some $j \in \{1, \ldots, r\}$ and some $G \in \Xi_j$. Then $G \leqslant G^*$, which contradicts the Pareto-minimality of G^*. Suppose G^* is not Pareto-minimal. Then there exists $G \in \Xi$ such that $G \leqslant G^*$. Therefore, for some $j \in \{1,\ldots,r\}$, $G_j < G_j^*$ and $G_i \leqq G_i^* \ \forall i \in \{1,\ldots,r\}$, $i \neq j$. But this contradicts $G_j^* \leqq G_j \ \forall G \in \Xi_j$. ■

It should be noted that the technique illustrated in Theorem 3.5, of minimizing one criterion subject to constraints on the other criteria, is a widely used approach to multicriteria decision problems.

Example 3.6 Consider the two-criteria problem with cost components

$$G_1(\cdot) = \sin y$$

$$G_2(\cdot) = \cos y$$

where $y \in [0, 2\pi]$ (see Figure 3.10*a*). In applying Theorem 3.5, we

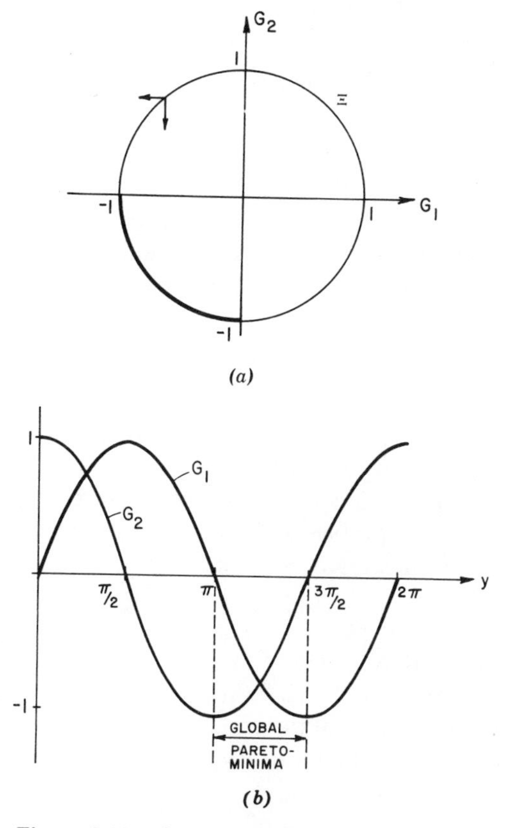

Figure 3.10. Cost functions for Example 3.6.

consider the following two problems of finding $y^* \in [0, 2\pi]$ to:

1. Minimize $\sin y$, subject to $\cos y^* - \cos y \geqq 0$.
2. Minimize $\cos y$, subject to $\sin y^* - \sin y \geqq 0$.

Thus we examine

$$L_1 = \sin y - \mu_1(\cos y^* - \cos y)$$

$$L_2 = \cos y - \mu_2(\sin y^* - \sin y)$$

and set

$$0 = \left.\frac{\partial L_1}{\partial y}\right|_{y^*} = \cos y^* - \mu_1 \sin y^*$$

$$0 = \left.\frac{\partial L_2}{\partial y}\right|_{y^*} = -\sin y^* + \mu_2 \cos y^*$$

where $\mu_1 \geqq 0$, $\mu_2 \geqq 0$. Thus we have

$$0 \leqq \mu_1 = \cot y^*$$

$$0 \leqq \mu_2 = \tan y^*$$

and the above first-order necessary conditions yield the following candidate Pareto-minimal points:

$$0 \leqq y^* \leqq \frac{\pi}{2}$$

$$\pi \leqq y^* \leqq \frac{3\pi}{2}.$$

By examining the cost functions over these two regions, we find that the global solutions to problems **1** and **2** corresponding to the region $\pi \leqq y^* \leqq 3\pi/2$, as illustrated in Figure 3.10*b*.

3.7 SCALARIZATION

Pareto-minimal points for multiple-criteria optimization problems may often be obtained by seeking minima of a scalar-valued "weighted sum" of the form

$$\eta^T G = \eta_1 G_1 + \cdots + \eta_r G_r$$

where $G=[G_1,\ldots,G_r]^T$ is the cost vector and $\eta=[\eta_1,\ldots,\eta_r]^T$ is a nonzero vector of nonnegative weights. This approach, as illustrated in Figure 3.11, is valid for a large class of problems, but the approach is not valid in all cases. The necessary conditions of Theorem 3.3 are the same as those for a local minimum of $\eta^T G$ with respect to y, but a local Pareto-minimal point may actually correspond, for example, to a local maximum of $\eta^T G$.

We have the following sufficiency condition for the scalarization method.

Theorem 3.6 If $G^* \in \Xi$ and $\eta \in E^r$ satisfy either

(i) $\eta^T G^* \leqq \eta^T G \qquad \forall G \in \Xi$, with $\eta > 0$

or

(ii) $\eta^T G^* < \eta^T G \qquad \forall G \in \Xi,\ G \neq G^*$, with $\eta \geqslant 0$,

then G^* is a Pareto-minimal solution.

> ***Proof*** Suppose G^* is not Pareto-minimal. Then, from Definition 3.1, there exists a vector $G \in \Xi$ such that $G \leqslant G^*$. For (i) $\eta > 0$ implies $\eta^T G < \eta^T G^*$, which would contradict $\eta^T G^* \leqq \eta^T G\ \forall G \in \Xi$. For (ii) $\eta \geqslant 0$ implies $\eta^T G \leqq \eta^T G^*$, which would contradict $\eta^T G^* < \eta^T G\ \forall G \in \Xi,\ G \neq G^*$. ■

Theorem 3.6 applies to an arbitrary cost set Ξ, and is a sufficiency condition for Pareto-minimality. A necessary and sufficient condition can be established for the case where Ξ satisfies a convexity condition. As a prelude, we first present some results on convex sets.

A set $A \subseteq E^r$ is *convex* if and only if for each pair of vectors z^1, z^2 in A and for each scalar $\alpha \in [0,1]$, $\alpha z^1 + (1-\alpha) z^2 \in A$.

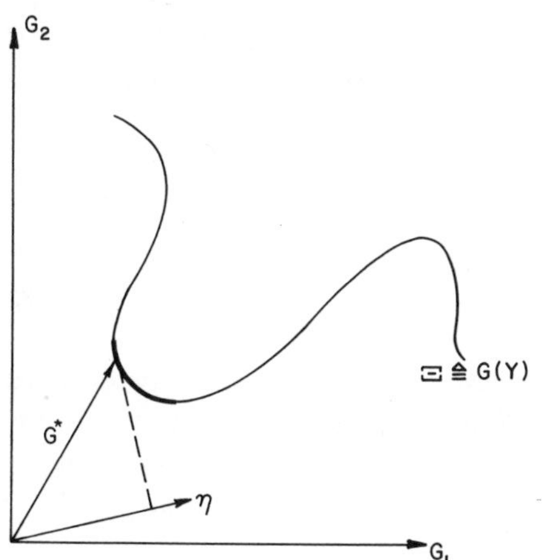

Figure 3.11 Scalarization by weighted sum.

A *hyperplane* H in E^r is an $(r-1)$-dimensional manifold of the form

$$H=\{z\in E^r \mid \eta^T z=b\}$$

where η is a nonzero vector and b is a scalar. The hyperplane H divides E^r into two open half-spaces

$$H^+=\{z\in E^r \mid \eta^T z>b\}$$

$$H^-=\{z\in E^r \mid \eta^T z<b\}$$

and forms the common boundary of H^+ and H^-. The sets H, H^+, and H^- are mutually disjoint and $H\cup H^+\cup H^-=E^r$.

A *support hyperplane* for a convex set A is a hyperplane H through a boundary point of A such that either $A\subseteq H\cup H^+$ or $A\subseteq H\cup H^-$; that is, all points of A are contained in one of the closed half-spaces defined by H.

The following lemma is a fundamental property of convex sets and may be taken as an alternate definition of convexity.

Lemma 3.8 A set $A\subseteq E^r$ is convex if and only if through every boundary point of A there exists at least one support hyperplane for A.

Proof See Rockafellar (1970). ■

Denoting the interior of a set A by $\mathring{A}$, another fundamental property of convex sets is given by the following lemma.

Lemma 3.9 If A and B are two nonempty convex sets with $\mathring{A}\cap\mathring{B}=\varnothing$, then there exists at least one *separating hyperplane* H such that $A\subseteq H\cup H^+$ and $B\subseteq H\cup H^-$. In other words, there exists a nonzero vector η and a scalar b such that

(i) $\eta^T z\geqq b \qquad \forall z\in A$

(ii) $\eta^T z\leqq b \qquad \forall z\in B$.

Proof See Rockafellar (1970). ■

Figure 3.12 is an illustration of a separating hyperplane for two convex sets in E^2. Note that the restriction in Lemma 3.9 is that the interiors of the two sets A and B must be disjoint, although the sets may have boundary points in common.

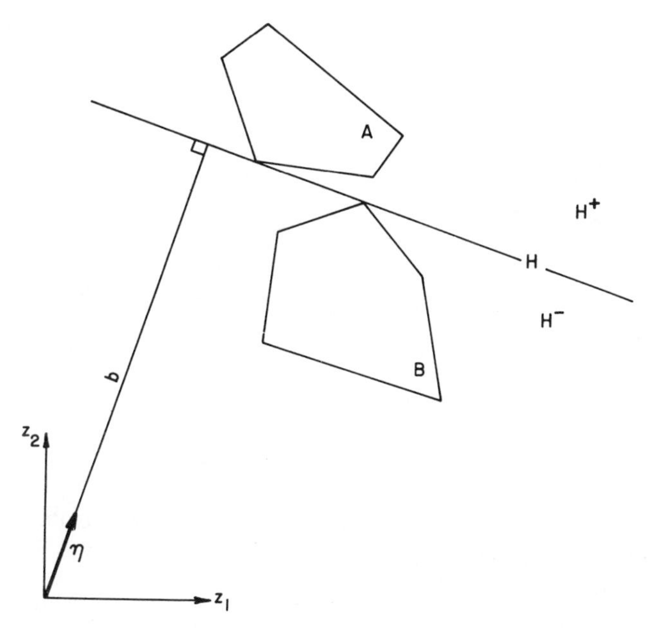

Figure 3.12. A separating hyperplane in E^2.

Lemmas 3.8 and 3.9 provide a relatively simple means for establishing a necessary and sufficient condition for Pareto-minima similar to the sufficiency condition of Theorem 3.6 providing the set Ξ is convex.

Theorem 3.7 If Ξ is convex, then $G^* \in \Xi$ is a Pareto-minimal point if and only if there exists a vector $\eta \geqslant 0$ such that

$$\eta^T G^* \leqq \eta^T G \qquad \forall G \in \Xi.$$

Proof From the contact theorem, G^* is Pareto-minimal if and only if

$$C(G^*) \cap \Xi = \{G^*\}$$

where $C(G^*)$ is the closed negative orthant centered at G^*:

$$C(G^*) = \{G \in E^r | G = G^* + \mu,\ \mu \leqq 0\}.$$

Hence

$$\hat{C}(G^*) \cap \Xi = \varnothing$$

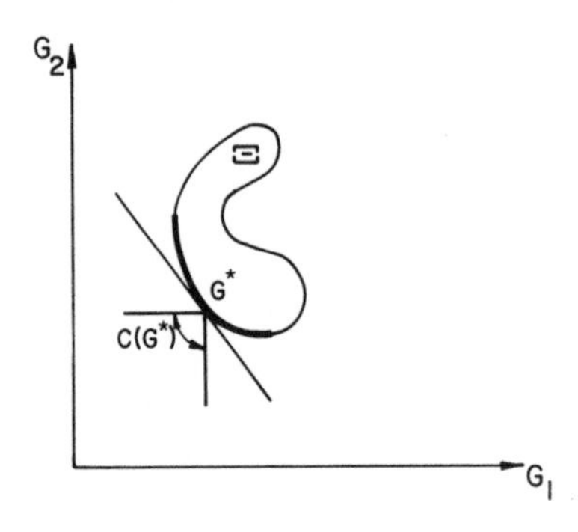

Figure 3.13. An illustration of the geometry for Theorem 3.7.

where

$$\hat{C}(G^*)=\{G\in E^r|G=G^*+\mu,\ \mu\leqslant 0\}.$$

Note that $\hat{C}(G^*)$ and Ξ are disjoint and convex. Hence from Lemma 3.9 there exists a separating hyperplane between $\hat{C}(G^*)$ and Ξ, that is, there exists a vector $\eta\geqslant 0$ such that

$$\eta^T G^* \leqq \eta^T G \qquad \forall G\in\Xi.$$

■

As illustrated in Figure 3.13, Theorem 3.7 is a statement of the existence of a separating hyperplane between the cost set Ξ and the closed negative orthant $C(G^*)$ at a Pareto-minimal point G^* providing Ξ is convex. Figure 3.14 illustrates a case where no such separating hyperplane exists at G^* even though G^* is Pareto-minimal (based on the contact theorem).

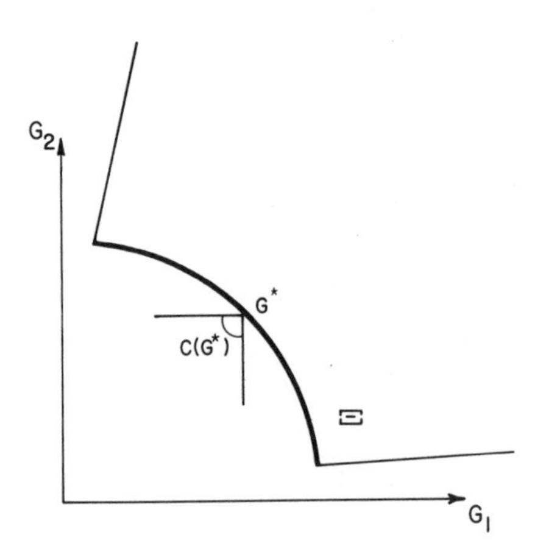

Figure 3.14 A Pareto-minimal point with no separating hyperplane.

The assumption of convexity in Theorem 3.7 is overly restrictive, since the existence of support hyperplanes at (boundary) points of Ξ is needed only at the Pareto-minimal points. See Lin (1976) for a development of the corresponding necessary condition using the concept of directional convexity (Holtzman and Halkin, 1966).

3.8 EXERCISES

3.1 Determine local Pareto-minimal point candidates for the vector cost function $G(y)=[y_1^2+y_2^2,(y_1-1)^2+(y_2-1)^2]^T$ where $y\in E^2$.

3.2 Determine local Pareto-minimal point candidates for the vector cost function $G(y)=[\frac{1}{3}y^3-\frac{5}{2}y^2+4y,-\frac{1}{2}(y-3)^2]^T$ where $y\in Y=\{y\in E^1 | y\geqq 0 \text{ and } 10-y\geqq 0\}$.

3.3 Determine local Pareto-minimal point candidates for the vector cost $G(y)=[y_2, y_1y_2]^T$ where $y\in Y=\{y\in E^2 | y_1+y_2=10\}$.

3.4 Determine local Pareto-minimal point candidates for the vector cost function $G(y)=[y_1^2+y_2^2,(y_1-2)^2+y_2^2]^T$ where $y\in Y$ with Y given by $Y=\{y\in E^2 | 1+y_2-y_1\geqq 0\}$.

3.5 Determine local Pareto-minimal point candidates for the vector cost function $G(y)=[y_1^2+y_2^2-y_1y_2, 2(y_1-1)^2+y_2^2, y_1^2+2(y_2-1)^2]^T$ where $y\in E^2$.

3.6 Let x_1=biomass of prey, x_2=biomass of predators. It has been proposed that the dynamics of a prey-predator system may be described by

$$\dot{x}_1=x_1(u_1-x_2)$$

$$\dot{x}_2=x_2(x_1-u_2)$$

where u_1 and u_2 are nonnegative "control" parameters provided by nature. Under steady-state equilibrium conditions ($\dot{x}_1=\dot{x}_2=0$), nature allows for a maximum amount of total biomass, that is

$$x_1+x_2\leqq B.$$

Consider the equilibrium situation only. The first species maximizes its growth term. The second species maximizes its biomass. Thus

$$H_1(\cdot)=x_1u_1$$

$$H_2(\cdot)=x_2.$$

Determine the set of Pareto-minimal values for u_1 and u_2.

3.7 The equilibrium (steady-state) nondimensional populations x_1 and x_2 of two species of fish in a lake may be modeled as

$$2-2x_1-x_2-2u=0$$

$$x_1-x_2-u=0$$

where $u\geqq 0$ is the fishing level. Two groups of fishermen would like to have the populations at $(x_1, x_2)=(1,0)$ and $(x_1, x_2)=(0,1)$, respectively. Since neither of these are possible, the two groups, respectively, seek

$$\min_u \left\{G_1(\cdot)=(x_1-1)^2+x_2^2\right\}$$

$$\min_u \left\{G_2(\cdot)=x_1^2+(x_2-1)^2\right\}.$$

Determine the set of Pareto-minimal fishing levels u^* where $u\geqq 0$.

3.8 A mechanical spring-mass-dashpot system is acted upon by two controllers plus a bias force as follows:

$$\ddot{y}+0.5\dot{y}+y=u^2+\dot{v}+f$$

where y=displacement of mass, u=input of first controller, v=input of second controller, and f=0.5 (bias force). The dot denotes differentiation with respect to time. The above second-order system is equivalent to the first-order systems

$$\dot{y}_1=-0.5y_1+y_2+v$$

$$\dot{y}_2=-y_1+0.5+u^2$$

where $y_1=y$. The second controller uses feedback control

$$v=-y_3y_1$$

where the constant of proportionality y_3 is restricted by

$$0\leqq y_3\leqq 1.$$

The first controller uses constant control

$$u=y_4,$$

which is restricted by

$$0 \leqq y_4 \leqq 1.$$

With y_3 and y_4 constant, there exist equilibrium solutions for y_1 and y_2 given by

$$y_2 = (y_3 + 0.5) y_1$$

$$y_1 = y_4^2 + 0.5.$$

The objective is to choose y_3 and y_4 to minimize the following cost functions at equilibrium:

$$G_1(\cdot) = y_1^2 + y_2^2$$

$$G_2(\cdot) = -(y_1 + y_2).$$

Determine values of y_3 and y_4 that yield candidate Pareto-minimal solutions for the vector-valued cost with components $G_1(\cdot)$ and $G_2(\cdot)$.

3.9 Show that the two statements of the theorem of the alternative, Theorems 1.6 and Theorems 3.2, are equivalent. *Hint*: Use the cones defined by (3.14)–(3.18) and Lemma 2.5.

Chapter Four

CONTROL NOTATION

4.1. INTRODUCTION

Consider again the scalar minimization problem of determining $y^* \in Y \subseteq E^m$ to minimize $G(\cdot): Y \to E^1$. The constraint set Y is defined by a system of n equality constraints

$$g(y)=0 \tag{4.1}$$

and q inequality constraints

$$h(y) \geqq 0, \tag{4.2}$$

where $g(\cdot): E^m \to E^n$ and $h(\cdot): E^m \to E^q$ are assumed C^1.

If the equality constraints (4.1) are independent, then there are $m-n$ degrees of freedom for choosing the m components of y at internal points of Y [where $h(y)>0$]. The number of degrees of freedom is further reduced at points where one or more of the inequality constraints become active. However, the fact that the equality constraints must be satisfied throughout Y suggests using the equality constraints to solve for n of the variables $y_1, \ldots, y_m$ as explicit functions of the remaining $m-n$ variables. If the resulting n functions are substituted into $G(\cdot)$ and $h(\cdot)$, the dimension of the problem is reduced from m to $m-n$. A reduction in dimension results in obvious computational advantages.

Such a procedure, called the method of direct substitution, often fails because of the difficulty in obtaining the required explicit functional relationships. Fortunately, however, the idea behind the method of direct substitution can still be used to advantage if it is possible to implicitly solve the equality constraints for n of the variables as functions of the remaining $m-n$ variables. The method in this case, which could be called the method of implicit substitution, requires a description of the problem in terms of a new notation, called control notation.

In control notation the variables $y=[y_1,\ldots,y_m]^T$ are relabeled so that the degrees of freedom at internal points of Y are explicitly identified. In particular, the components of y are relabeled to form a *state vector* $x=[x_1,\ldots,x_n]^T$ and a *control vector* $u=[u_1,\ldots,u_s]^T$, $s=m-n$, such that

$$y^T=[x^T,u^T] \tag{4.3}$$

where the components of y have been reordered conceptually to group all of the state variables together and all of the control variables together.

Loosely speaking, the $s=m-n$ control variables u may be thought of as those components of y that may be freely chosen (locally) at internal points of Y. The n state variables x are then determined from the n equality constraints

$$g(x,u)=0 \tag{4.4}$$

where $g(\cdot): E^n\times E^s\to E^n$. Thus a component of u is simply one of the components of the y vector that has been identified with a degree of freedom.

The selection process for state and control variables is not unique, but the implication is that, given all components of the control vector u, the state vector x can be determined from the equality constraints (4.4). Thus from the implicit function theorem, we require that at a local minimal point $(x^*,u^*)\in Y$

$$\left|\frac{\partial g(x^*,u^*)}{\partial x}\right|\neq 0, \tag{4.5}$$

where $\partial g/\partial x$ is the $n\times n$ matrix with elements $\partial g_i/\partial x_j$, i, $j=1,\ldots,n$. This requirement may indeed influence the choice of the variables u and x.

Example 4.1 Consider the following case with two equality constraints in four variables:

$$g_1(\cdot)=y_1-y_2-y_3^3+y_4^2+4=0$$

$$g_2(\cdot)=2y_1+y_2-2y_3^2+3y_4^4+8=0.$$

The selection $x=(x_1,x_2)\triangleq(y_1,y_2)$ and $u=(u_1,u_2)\triangleq(y_3,y_4)$ produces

$$\left|\frac{\partial g}{\partial x}\right|=\begin{vmatrix}1 & -1\\ 2 & 1\end{vmatrix}=3\neq 0,$$

which satisfies (4.5) everywhere. Alternately, the selection $x=(x_1, x_2) \triangleq (y_3, y_4)$ and $u=(u_1, u_2) \triangleq (y_1, y_2)$ produces

$$\left|\frac{\partial g}{\partial x}\right| = \begin{vmatrix} -3x_1^2 & 2x_2 \\ -4x_1 & 12x_2^3 \end{vmatrix}$$
$$= -4x_1x_2(9x_1x_2^2-2),$$

which does not satisfy (4.5) on the line $x_1=0$, on the line $x_2=0$, and on the curve $9x_1x_2^2-2=0$.

4.2 FIRST-ORDER NECESSARY CONDITIONS

Use of control notation allows the control u to be chosen independently of the state x at internal points of Y. However, at noninternal points of Y we cannot select the control independently of the state if the inequality constraints (4.2) depend on x. This is simply because a pair $(x, u)=y$ satisfying (4.1) may not satisfy (4.2). One of the objectives of control notation is to have the inequality constraints (4.2) independent of the state x, that is, $h(u) \geqq 0$. Whenever possible, the state and control variables should be selected to achieve this objective. Only in this case can the control be chosen independently of the state for all points of the set Y. In other words, given a control u satisfying $h(u) \geqq 0$, we can think of the state as being determined implicitly by (4.4) provided (4.4) has a solution. The general case considered in the following is subsequently specialized to the state independent inequality constraints case. As might be expected, the latter case results in simpler necessary conditions of reduced dimension.

Suppose that the y vector has been relabeled in terms of state and control vectors x and u, respectively. Then the scalar minimization problem is to determine u^* to

$$\text{minimize } G(x, u)$$

subject to

$$g(x, u)=0 \tag{4.6}$$

$$h(x, u) \geqq 0 \tag{4.7}$$

where $G(\cdot): E^n \times E^s \to E^1$, $g(\cdot): E^n \times E^s \to E^n$, and $h(\cdot): E^n \times E^s \to E^q$

are assumed C^1 in all of their arguments, and

$$\left|\frac{\partial g(x^*, u^*)}{\partial x}\right| \neq 0 \tag{4.8}$$

where x^* is the solution to (4.6) corresponding to $u=u^*$. At this point first-order necessary conditions could be written down directly by use of Theorem 2.3. However, it is useful to rederive them in terms of the new notation.

For the remainder of this text we assume (4.8) is satisfied. The implicit function theorem guarantees the existence of a C^1 function $\xi(\cdot): E^s \to E^n$ and of a ball B about $u^* \in E^s$ such that, for all $u \in B$, the solution x to

$$g(x, u)=0 \tag{4.9}$$

is given by

$$x=\xi(u). \tag{4.10}$$

Thus if we define

$$\tilde{G}(u)=G(\xi(u), u) \tag{4.11}$$

$$\tilde{g}(u)=g(\xi(u), u) \tag{4.12}$$

$$\tilde{h}(u)=h(\xi(u), u), \tag{4.13}$$

an equivalent problem may be posed. Note that since

$$\tilde{g}(u) \equiv 0 \tag{4.14}$$

we have no equality constraints. The equivalent problem becomes one of determining u^* to

$$\text{minimize } \tilde{G}(u)$$

subject to

$$\tilde{h}(u) \geqq 0.$$

Let

$$U=\left\{u \in E^s \mid \tilde{h}(u) \geqq 0\right\} \tag{4.15}$$

and

$$T=\{e\in E^s \mid e \text{ tangent to } U \text{ at } u^*\}$$

be the control constraint set and its tangent cone at u^*, respectively. Then from Lemma 2.1, if u^* is a local minimal point for $\tilde{G}(\cdot):U\to E^1$, which is C^1 on U, it is necessary that

$$\frac{\partial \tilde{G}(u^*)}{\partial u}e \geqq 0 \tag{4.16}$$

for all $e\in T$. Furthermore, the condition

$$\frac{\partial \tilde{G}(u^*)}{\partial u}e > 0 \tag{4.17}$$

for all nonzero $e\in T$ is sufficient for a proper local minimum at u^*. We now assume that U is regular so that the tangent cone T to U at u^* consists of exactly those vectors e that satisfy

$$\frac{\partial \tilde{h}_j(u^*)}{\partial u}e \geqq 0 \qquad \forall j\in Q^* \tag{4.18}$$

where

$$Q^*=\{j \mid \tilde{h}_j(u^*)=0\}=\{1,\dots,q^*\}$$

denotes the active inequality constraints at u^*. If Q^* is empty, then $T=E^s$. Because of the identity (4.14) it follows that

$$\frac{\partial \tilde{g}(u^*)}{\partial u}=\frac{\partial g(x^*,u^*)}{\partial x}\frac{\partial \xi(u^*)}{\partial u}+\frac{\partial g(x^*,u^*)}{\partial u}=0. \tag{4.19}$$

Thus

$$\frac{\partial \xi(u^*)}{\partial u}=-\left[\frac{\partial g(x^*,u^*)}{\partial x}\right]^{-1}\frac{\partial g(x^*,u^*)}{\partial u} \tag{4.20}$$

and (4.16) and (4.18) may be written as

$$\left[\frac{\partial G(x^*,u^*)}{\partial u}-\frac{\partial G(x^*,u^*)}{\partial x}\left[\frac{\partial g(x^*,u^*)}{\partial x}\right]^{-1}\frac{\partial g(x^*,u^*)}{\partial u}\right]e\geqq 0 \tag{4.21}$$

and

$$\left[\frac{\partial h_j(x^*,u^*)}{\partial u}-\frac{\partial h_j(x^*,u^*)}{\partial x}\left[\frac{\partial g(x^*,u^*)}{\partial x}\right]^{-1}\frac{\partial g(x^*,u^*)}{\partial u}\right]e\geqq 0 \qquad \forall j\in Q^*. \tag{4.22}$$

In terms of the state x and control u we have the following theorem.

Theorem 4.1 If $u^*\in U$ is a regular local minimum point for $G(\cdot)$ and $x^*=\xi(u^*)$ is the solution to $g(x,u^*)=0$ then there exists a vector $\nu\in E^n$ defined by

$$\frac{\partial J(x^*,u^*,\nu)}{\partial x}=0 \tag{4.23}$$

and vectors $\rho\in E^n$ and $\mu\in E^q$, such that

$$\frac{\partial V(x^*,u^*,\rho,\mu)}{\partial x}=0 \tag{4.24}$$

$$\frac{\partial L(x^*,u^*,\nu,\rho,\mu)}{\partial u}=0 \tag{4.25}$$

$$g(x^*,u^*)=0 \tag{4.26}$$

$$\mu\geqq 0 \tag{4.27}$$

$$h(x^*,u^*)\geqq 0 \tag{4.28}$$

$$\mu^T h(x^*,u^*)=0 \tag{4.29}$$

where

$$J(x,u,\nu)\triangleq G(x,u)-\nu^T g(x,u) \tag{4.30}$$

$$V(x,u,\rho,\mu)\triangleq \mu^T h(x,u)+\rho^T g(x,u) \tag{4.31}$$

$$L(x,u,\nu,\rho,\mu)\triangleq J-V. \tag{4.32}$$

Proof If we define the vector $\nu\in E^n$ as

$$\nu^T\triangleq\frac{\partial G(x^*,u^*)}{\partial x}\left(\frac{\partial g(x^*,u^*)}{\partial x}\right)^{-1}, \tag{4.33}$$

then (4.21) and (4.22) may be written as

$$\left[\frac{\partial G}{\partial u} - \nu^T \frac{\partial g}{\partial u}\right] e \geqq 0$$

for all $e \in E^n$ such that

$$\left[\frac{\partial \hat{h}}{\partial u} - \frac{\partial \hat{h}}{\partial x}\left[\frac{\partial g}{\partial x}\right]^{-1} \frac{\partial g}{\partial u}\right] e \geqq 0$$

where all quantities are evaluated at (x^*, u^*) and where $\hat{h}(\cdot) = [h_1(\cdot), \ldots, h_{q^*}(\cdot)]^T$ denotes the active inequality constraints at (x^*, u^*). It follows from Farkas' lemma that there exists a nonnegative multiplier vector $\mu \in E^q$, $\mu \geqq 0$, such that at (x^*, u^*)

$$\frac{\partial G}{\partial u} - \nu^T \frac{\partial g}{\partial u} = \mu^T \left\{ \frac{\partial h}{\partial u} - \frac{\partial h}{\partial x}\left[\frac{\partial g}{\partial x}\right]^{-1} \frac{\partial g}{\partial h} \right\} \tag{4.34}$$

where $\mu_j h_j(x^*, u^*) = 0$ for all $j \in Q$. Under the requirements (4.26)–(4.29), with the definitions (4.30)–(4.32) and the definition

$$\rho^T \triangleq -\mu^T \frac{\partial h}{\partial x}\left[\frac{\partial g}{\partial x}\right]^{-1} \tag{4.35}$$

at (x^*, u^*), (4.34) is equivalent to (4.25), (4.33) is equivalent to (4.23), and (4.35) is equivalent to (4.24). ■

Note that, if the control variables can be chosen so that the inequality constraints are functions of u only (i.e., $h(u) \equiv \tilde{h}(u)$), then from (4.8), (4.24), and (4.31) the vector ρ is the zero vector and the necessary conditions are greatly simplified.

Corollary 4.1 Let $h(\cdot)$ be a function of u only. If $u^* \in U$ is a regular local minimum point for $G(\cdot)$ and $x^* = \xi(u^*)$ is the solution to $g(x, u^*) = 0$ then there exists a vector $\lambda \in E^n$ defined by

$$\frac{\partial L(x^*, u^*, \lambda, \mu)}{\partial x} = 0 \tag{4.36}$$

and a vector $\mu \in E^q$ such that

$$\frac{\partial L(x^*, u^*, \lambda, u)}{\partial u} = 0 \tag{4.37}$$

$$g(x^*, u^*) = 0$$

$$\mu \geqq 0$$

$$h(u^*) \geqq 0$$

$$\mu^T h(u^*) = 0$$

where

$$L(x, u, \lambda, \mu) = G(x, u) - \lambda^T g(x, u) - \mu^T h(u). \tag{4.38}$$

Note that Corollary 4.1 is stated in terms of a single scalar function $L(\cdot)$ instead of the three functions $L(\cdot)$, $V(\cdot)$, and $J(\cdot)$ employed in Theorem 4.1. This simplification is a direct result of the inequality constraints being independent of x. In such a situation (4.24) implies $\rho = 0$ and, in addition, (4.23) is equivalent to (4.36) since $\partial h / \partial x \equiv 0$.

Note that we have used ν in Theorem 4.1 and λ in the corollary. The reason for this is that, if Theorem 2.3 were applied directly to the problem

$$\text{minimize } G(x, u)$$

subject to

$$g(x, u) = 0$$

$$h(u) \geqq 0,$$

we would have

$$L(x, u, \lambda, \mu) = G(x, u) - \lambda^T g(x, u) - \mu^T h(u)$$

with the necessary conditions

$$\frac{\partial L}{\partial x} = \frac{\partial G}{\partial x} - \lambda^T \frac{\partial g}{\partial x} = 0$$

$$\frac{\partial L}{\partial u} = \frac{\partial G}{\partial u} - \lambda^T \frac{\partial g}{\partial u} - \mu^T \frac{\partial h}{\partial u} = 0$$

being identical to (4.36) and (4.37). However, a direct application of Theorem 2.3 to the case where $h(\cdot)$ is a function of both x and u would result in λ being a function of μ. Details are left as a homework exercise. The ν used in Theorem 4.1 is not, in general, the λ of Theorem 2.3.

The control notation represents a conceptual change in viewpoint for solving problems. Instead of seeking an m-dimensional vector $y^* \in Y$ that minimizes the cost function $G(y)=G(x,u)$, the problem has been transformed, through the relation $x=\xi(u)$ implicit in $g(x,u)=0$, to one in which we seek an s-dimensional vector $u^* \in U$ that minimizes the cost function $G(y)=G(x,u)=G[\xi(u),u]$.

Example 4.2 Minimize

$$G=-y_1y_2y_3$$

subject to

$$g=y_1^2+2y_2^2+3y_3^2-27=0$$

$$h_1=y_1-1\geqq 0$$

$$h_2=y_2-1\geqq 0$$

$$h_3=5-y_2\geqq 0$$

$$h_4=y_3-1\geqq 0$$

$$h_5=5-y_3\geqq 0.$$

Using control notation, let

$$x=y_1$$

$$u_1=y_2$$

$$u_2=y_3.$$

In terms of this notation the problem becomes one of minimizing

$$G=-xu_1u_2$$

subject to

$$g=x^2+2u_1^2+3u_2^2-27=0$$

$$h_1=\sqrt{27-2u_1^2-3u_2^2}-1\geqq 0$$

$$h_2 = u_1 - 1 \geqq 0$$
$$h_3 = 5 - u_1 \geqq 0$$
$$h_4 = u_2 - 1 \geqq 0$$
$$h_5 = 5 - u_2 \geqq 0.$$

Note that $|\partial g/\partial x| = 2x \neq 0$ since $x \geqq 1$. Since $h(\cdot)$ is a function of u only we use Corollary 4.1. Let

$$L = -xu_1u_2 - \lambda(x^2 + 2u_1^2 + 3u_2^2 - 27)$$
$$-\mu_1\left(\sqrt{27 - 2u_1^2 - 3u_2^2} - 1\right) - \mu_2(u_1 - 1) - \mu_3(5 - u_1)$$
$$-\mu_4(u_2 - 1) - \mu_5(5 - u_2)$$

so that (4.36) and (4.37) become

$$\frac{\partial L}{\partial x} = 0 = -u_1u_2 - 2\lambda x$$
$$\frac{\partial L}{\partial u_1} = 0 = -xu_2 - 4\lambda u_1 + \frac{2u_1\mu_1}{x} - \mu_2 + \mu_3$$
$$\frac{\partial L}{\partial u_2} = 0 = -xu_1 - 6\lambda u_2 + \frac{3u_2\mu_1}{x} - \mu_4 + \mu_5.$$

From $\dfrac{\partial L}{\partial x} = 0$, since $x \geqq 1$,

$$\lambda = -\frac{u_1u_2}{2x}.$$

Substituting for λ and using $g = 0$ the latter two necessary conditions become

$$0 = 4u_1^2u_2 + 3u_2^3 - 27u_2 + 2u_1\mu_1 + (\mu_3 - \mu_2)x$$
$$0 = 2u_1^3 + 6u_1u_2^2 - 27u_1 + 3u_2\mu_1 + (\mu_5 - \mu_4)x.$$

We check for internal solutions by setting $\mu_i = 0, i = 1, \ldots, 5$ to obtain

$$0 = u_2(4u_1^2 + 3u_2^2 - 27)$$
$$0 = u_1(2u_1^2 + 6u_2^2 - 27).$$

These equations along with $g=0$ imply

$$x=\pm 3,\ u_1=\frac{3}{\sqrt{2}},\ u_2=\sqrt{3}.$$

Other candidates are obtained by examining the necessary conditions with one or more of the inequality constraints active. Note that neither $h_3=0$ ($u_1=5$) nor $h_5=0$ ($u_2=5$) satisfy $u\in U$, hence they need not be checked. The remaining three inequality constraints yield no further candidates. For example if only h_1 is active so that $x=1$, $\mu_1\geqq 0$, $\mu_2=\mu_3=\mu_4=\mu_5=0$, then the above necessary conditions become

$$4u_1^2u_2+3u_2^3-27u_2+2u_1\mu_1=0$$
$$2u_1^3+6u_1u_2^2-27u_1+3u_2\mu_1=0.$$

In addition from $g=0$ with $x=1$ we obtain

$$(2u_1^2-1)u_2+2u_1\mu_1=0$$
$$(3u_2^2-1)u_1+3u_2\mu_1=0.$$

Both of these equations yield $\mu_1<0$ for $u_1\geqq 1$ and $u_2\geqq 1$ hence no additional candidate is obtained.

4.3 LAGRANGE MULTIPLIERS AS SENSITIVITY COEFFICIENTS

Let (x^*, u^*) be a regular local minimal point for $G(x,u)$ subject to (4.6) and (4.7). Let ν, ρ, and μ be the Lagrange multiplier vectors satisfying the necessary conditions of Theorem 4.1. Let $\hat{h}(x^*,u^*)$ denote the active inequality constraints at (x^*,u^*). One is frequently interested in determining how sensitive $G(x,u)$ is to changes in x and u in a neighborhood of (x^*,u^*). Let Δx and Δu represent arbitrary small changes in the solution (x^*,u^*) where, in general, $(x,u)=(x^*+\Delta x, u^*+\Delta u)$ may violate the equality constraints or the active inequality constraints at (x^*,u^*). The changes Δx and Δu are taken sufficiently small, however, so that no other inequality constraints become active. Let ΔG, Δg, and $\Delta\hat{h}$ be the corresponding changes in the cost and constraints. To first-order we have

$$\Delta G=\frac{\partial G}{\partial x}\Delta x+\frac{\partial G}{\partial u}\Delta u \tag{4.39}$$

$$\Delta g=\frac{\partial g}{\partial x}\Delta x+\frac{\partial g}{\partial u}\Delta u \tag{4.40}$$

$$\Delta\hat{h}=\frac{\partial\hat{h}}{\partial x}\Delta x+\frac{\partial\hat{h}}{\partial u}\Delta u \tag{4.41}$$

where all partial derivatives are evaluated at (x^*, u^*). From (4.23) and (4.24) we have

$$\nu^T = \frac{\partial G}{\partial x}\left[\frac{\partial g}{\partial x}\right]^{-1} \tag{4.42}$$

$$\rho^T = -\mu^T \frac{\partial h}{\partial x}\left[\frac{\partial g}{\partial x}\right]^{-1} \tag{4.43}$$

and from (4.25)

$$\frac{\partial J}{\partial u} = \mu^T \frac{\partial h}{\partial u} + \rho^T \frac{\partial g}{\partial u}. \tag{4.44}$$

Solving (4.40) for Δx and substituting into (4.39) and (4.41) and using (4.30), (4.42), and (4.43) yields

$$\Delta G = \nu^T \Delta g + \frac{\partial J}{\partial u}\Delta u$$

$$\mu^T \Delta h = -\rho^T \Delta g + \left(\mu^T \frac{\partial h}{\partial u} + \rho^T \frac{\partial g}{\partial u}\right)\Delta u.$$

From (4.44) it follows that the above equations may be combined to yield

$$\Delta G = (\nu + \rho)^T \Delta g + \mu^T \Delta h.$$

If (x^*, u^*) is an internal point, that is, if $h(x^*, u^*) > 0$, then $\mu = 0$, $\rho = 0$, and we have

$$\Delta G = \nu^T \Delta g,$$

that is, ν indicates the sensitivity of the cost to changes in the equality constraints at an internal minimal point. If the perturbations $(\Delta x, \Delta u)$ are such that the equality constraints remain satisfied, that is, if $\Delta g = 0$, then

$$\Delta G = \mu^T \Delta h.$$

Thus μ indicates the sensitivity of the cost to changes in the inequality constraints while maintaining the equality constraints. Finally, if the perturbations $(\Delta x, \Delta u)$ are such that the active inequality constraints remain active $(\Delta h = 0)$, then

$$\Delta G = (\nu + \rho)^T \Delta g.$$

Thus ρ indicates the (additional) sensitivity of the cost due to changes in the equality constraints, while maintaining the active inequality constraints. Note that if $\rho=0$ the sensitivity is the same as in the case where (x^*, u^*) is an internal minimal point. Either of the conditions $h(x^*, u^*)>0$ or $h(\cdot)=h(u)$ are sufficient for $\rho=0$. But other conditions, such as $\partial h(x^*, u^*)/\partial x=0$, also yield $\rho=0$.

4.4 SECOND-ORDER CONDITIONS

In this section we develop second-order necessary conditions and sufficient conditions for the case where the inequality constraints are inactive. We assume $G(\cdot)$ and $g(\cdot)$ are C^2 and that (4.5) holds at (x^*, u^*). As before, we define

$$\tilde{G}(u)=G(\xi(u), u).$$

A first order necessary condition for a regular local internal minimum of $\tilde{G}(\cdot)$ is that u^* be stationary. The second-order conditions involve simply the eigenvalues of the matrix $\partial^2\tilde{G}/\partial u^2$ evaluated at u^*. Nonnegative eigenvalues are necessary at a local minimal point and positive eigenvalues, along with $\partial\tilde{G}/\partial u=0$, are sufficient for a local minimal point.

Since $\xi(\cdot)$ is not known *a priori*, we seek a formulation of the optimality conditions in terms of known functions. From the definition of $\tilde{G}(\cdot)$ we have

$$\frac{\partial \tilde{G}(u^*)}{\partial u}=\frac{\partial G(x^*, u^*)}{\partial x}\frac{\partial \xi(u^*)}{\partial u}+\frac{\partial G(x^*, u^*)}{\partial u}$$

so that from (4.20)

$$\frac{\partial \tilde{G}(u^*)}{\partial u}=\frac{\partial G(x^*, u^*)}{\partial u}-\frac{\partial G(x^*, u^*)}{\partial x}\left[\frac{\partial g(x^*, u^*)}{\partial x}\right]^{-1}\frac{\partial g(x^*, u^*)}{\partial u}.$$

At (x^*, u^*) we have

$$\frac{\partial L(x^*, u^*, \lambda)}{\partial x}=0 \tag{4.45}$$

where

$$L(x, u, \lambda)=G(x, u)-\lambda^T g(x, u). \tag{4.46}$$

In this case $\lambda = \nu$ as given by (4.33) so that

$$\frac{\partial \tilde{G}(u^*)}{\partial u} = \frac{\partial L(x^*, u^*, \lambda)}{\partial u} = \frac{\partial G(x^*, u^*)}{\partial u} - \lambda^T \frac{\partial g(x^*, u^*)}{\partial u}. \quad (4.47)$$

Defining

$$\tilde{L}(u) = L[\xi(u), u, \tilde{\lambda}(u)]$$

where $\tilde{\lambda}(u)$ is the solution to

$$\frac{\partial L(\xi(u), u, \tilde{\lambda})}{\partial x} = 0, \quad (4.48)$$

we have

$$\frac{\partial^2 \tilde{G}(u^*)}{\partial u^2} = \frac{\partial^2 \tilde{L}(u^*)}{\partial u^2} = \frac{\partial^2 L}{\partial u^2} + \frac{\partial^2 L}{\partial x \partial u}\frac{\partial \xi}{\partial u} + \frac{\partial^2 L}{\partial \lambda \partial u}\frac{\partial \tilde{\lambda}}{\partial u} \quad (4.49)$$

where all quantities are evaluated at (x^*, u^*) in (4.49) and in the sequel. Note that

$$\frac{\partial^2 L}{\partial \lambda \partial u} = -\left[\frac{\partial g}{\partial u}\right]^T \quad (4.50)$$

and

$$\frac{\partial^2 L}{\partial \lambda \partial x} = -\left[\frac{\partial g}{\partial x}\right]^T \quad (4.51)$$

and that $\partial \tilde{\lambda}/\partial u$ may be determined from (4.48), which must be satisfied for all u, that is

$$\frac{\partial^2 L}{\partial u \partial x} + \frac{\partial^2 L}{\partial x^2}\frac{\partial \xi}{\partial u} + \frac{\partial^2 L}{\partial \lambda \partial x}\frac{\partial \tilde{\lambda}}{\partial u} = 0.$$

Thus, using (4.51), we have

$$\frac{\partial \tilde{\lambda}}{\partial u} = \left\{\left[\frac{\partial g}{\partial x}\right]^T\right\}^{-1}\left[\frac{\partial^2 L}{\partial u \partial x} + \frac{\partial^2 L}{\partial x^2}\frac{\partial \xi}{\partial u}\right].$$

This result along with (4.20) allows (4.49) to be written as

$$\frac{\partial^2 \tilde{G}}{\partial u^2}=\frac{\partial^2 L}{\partial u^2}-\frac{\partial^2 L}{\partial x\,\partial u}\left[\frac{\partial g}{\partial x}\right]^{-1}\left[\frac{\partial g}{\partial u}\right]$$

$$-\left[\frac{\partial g}{\partial u}\right]^T\left\{\left[\frac{\partial g}{\partial x}\right]^T\right\}^{-1}\left\{\frac{\partial^2 L}{\partial u\,\partial x}-\frac{\partial^2 L}{\partial x^2}\left[\frac{\partial g}{\partial x}\right]^{-1}\left[\frac{\partial g}{\partial u}\right]\right\}$$

or

$$\frac{\partial^2 \tilde{G}}{\partial u^2}=\frac{\partial^2 L}{\partial u^2}-\frac{\partial^2 L}{\partial x\,\partial u}\left[\frac{\partial g}{\partial x}\right]^{-1}\left[\frac{\partial g}{\partial u}\right]-\left[\frac{\partial g}{\partial u}\right]^T\left\{\left[\frac{\partial g}{\partial x}\right]^T\right\}^{-1}\frac{\partial^2 L}{\partial u\,\partial x}$$

$$+\left[\frac{\partial g}{\partial u}\right]^T\left\{\left[\frac{\partial g}{\partial x}\right]^T\right\}^{-1}\frac{\partial^2 L}{\partial x^2}\left[\frac{\partial g}{\partial x}\right]^{-1}\left[\frac{\partial g}{\partial u}\right]. \tag{4.52}$$

For an alternate derivation of this result see Bryson and Ho (1975) and Vincent and Cliff (1970).

The matrix given in (4.52) must be positive semi-definite at a regular local internal minimal point. If the matrix is positive definite at (x^*, u^*) and if the first-order necessary conditions of Theorem 4.1 are satisfied with $h(x^*, u^*)>0$, then (x^*, u^*) is a regular local internal minimal point.

4.5 THE GOLDEN FLEECE

Having developed necessary conditions and sufficient conditions in control notation, the reader familiar with optimal control theory might logically wonder whether there exists a minimum principle for static systems analogous to Pontryagin's maximum principle for dynamic systems (Pontryagin et al., 1962; Boltyanskii, 1971). The conjectured minimum principle is $L(x^*, u^*, \lambda)=\min_{u\in U} L(x^*, u, \lambda)$. Such a theorem would be very useful indeed. However, the conjecture is false in general.

The temptation of a static minimum principle reminds us of the enchanted songs of the Sirens in Greek mythology. Both Jason and Odysseus were lured toward shipwreck on the dangerous rocks surrounding the Sirens' island. Odysseus resisted by securing himself to a strong mast on his ship and plugging the ears of his men. Jason heard another, lovelier song. The lure of the static minimum principle can also be resisted successfully.

It follows from (4.21) that a necessary condition for (x^*, u^*) to minimize $G(x,u)$ subject to $g(x,u)=0$ where $|\partial g(x^*, u^*)/\partial x| \neq 0$ is given by

$$\frac{\partial L(x^*, u^*, \lambda)}{\partial u} e \geqq 0 \qquad \text{for all } e \in T \tag{4.53}$$

where T is the tangent cone to U at (x^*, u^*), $L(x,u,\lambda)=G(x,u)-\lambda^T g(x,u)$ and $\lambda = \nu$ as given by (4.33). However, the condition

$$\frac{\partial L(x^*, u^*, \lambda)}{\partial u} e > 0 \qquad \text{for all } e \in T,$$

along with $g(x^*, u^*)=0$ and $|\partial g(x^*, u^*)/\partial x| \neq 0$, is *not* sufficient for $G(x,u)$ to take on a proper local minimum. In other words, $L(x^*, u^*, \lambda)$ need not have a local minimum with respect to u at x^*, u^*. Indeed it may even have a local maximum!

The reason a static minimum principle does not exist is that a perturbation δu in u alone will not guarantee that $g(x^*, u^* + \delta u)=0$. In general, a corresponding perturbation δx must also occur. This result may be contrasted with the dynamic maximum principle where it is possible to make a perturbation in the control at an instant in time without making an immediate change in the state. It should also be noted that if a static maximum principle were to exist, then the second-order conditions would only involve eigenvalues of $\partial^2 L/\partial u^2$ rather than the more complicated matrix given by (4.52).

Equation (4.53) is a necessary (but not sufficient) condition for $L(x^*, u^*, \lambda)$ to have a local minimum at u^*. The following example illustrates that $L(x^*, u^*, \lambda)$ need not be minimized at u^* and may even have a local maximum at u^*. For additional discussion, see Vincent and Cliff (1970).

Example 4.3 For scalar x and u minimize

$$G(x,u) = x^2 + \frac{\sqrt{3}}{2} u$$

subject to

$$g(x,u) = x - \sin u = 0$$

with $0 < u < \pi$.

Using Corollary 4.1 and defining

$$L(x,u,\lambda)=x^2+\frac{\sqrt{3}}{2}u-\lambda(x-\sin u),$$

the first-order necessary conditions are

$$\frac{\partial L}{\partial x}=0=2x-\lambda$$

$$\frac{\partial L}{\partial u}=0=\frac{\sqrt{3}}{2}+\lambda\cos u.$$

Hence

$$\lambda=2x=2\sin u$$

and, therefore,

$$\sin 2u=-\frac{\sqrt{3}}{2}.$$

Thus for $0<u<\pi$ we have the two candidates

$$u=\frac{2\pi}{3},\qquad x=\frac{\sqrt{3}}{2},\qquad \lambda=\sqrt{3}$$

and

$$u=\frac{5\pi}{6},\qquad x=\frac{1}{2},\qquad \lambda=1.$$

From the second-order conditions (4.52), with $x=\xi(u)=\sin u$, we have

$$\begin{aligned}\frac{\partial^2\tilde{G}}{\partial u^2}&=\frac{\partial^2 L}{\partial u^2}+\frac{\partial^2 L}{\partial x\,\partial u}\left[\frac{\partial\xi}{\partial u}\right]+\left[\frac{\partial\xi}{\partial u}\right]^T\frac{\partial^2 L}{\partial u\,\partial x}+\left[\frac{\partial\xi}{\partial u}\right]^T\frac{\partial^2 L}{\partial x^2}\left[\frac{\partial\xi}{\partial u}\right]\\&=-\lambda\sin u+0+0+2\cos^2 u\\&=-2\sin^2 u+2\cos^2 u\\&=2\cos 2u.\end{aligned}$$

At $u=2\pi/3$

$$\frac{\partial^2 \tilde{G}}{\partial u^2}=2\cos\frac{4\pi}{3}=-1,$$

which implies a local maximum. At $u=5\pi/6$

$$\frac{\partial^2 \tilde{G}}{\partial u^2}=2\cos\frac{5\pi}{3}=1$$

so that

$$u^*=\frac{5\pi}{6}, \qquad x^*=\frac{1}{2}, \qquad \lambda=1$$

is the minimal point. But for fixed x^* and λ, $L(x^*, u, \lambda)$ takes on a local maximum at u^*, as shown from

$$\frac{\partial^2 L}{\partial u^2}=-\lambda \sin u^*$$

$$=-2\sin^2 u^*<0.$$

4.6 EXERCISES

4.1 Determine all local minima for the following problem using both the notation of Chapter 2 and control notation: $G(y)=y_2^3-3y_1y_2$ where $y\in Y=\{y\in E^2 \mid y_1-2y_2-5=0,\ 2y_1+5y_2-19\geqq 0,\ y_1\geqq 0,\ y_2\geqq 0\}$.

4.2 Determine all local minima for the following problem using both the notation of Chapter 2 and control notation: $G(y)=-y_1y_2y_3$ where $y\in Y=\{y\in E^3 \mid 2y_1y_2+2y_2y_3-1=0,\ 0.5\leqq y_2\leqq 1,\ 0.5\leqq y_3\leqq 1,\ y_1\geqq 0\}$.

4.3 Solve Exercise 2.7 using control notation.

4.4 Solve Exercise 2.15 using control notation.

4.5 Convert Exercise 3.8 to control notation by letting $x_1=y_1$, $x_2=y_2$, $u_1=y_3$, and $u_2=y_4$. For the equilibrium situation determine u_1 to minimize $G_1(\cdot)$ for a given $0\leqq u_2\leqq 1$ and determine u_2 to minimize $G_2(\cdot)$ for a given $0\leqq u_1\leqq 1$. Draw a sketch, in state space, of these solutions for all admissible u_1, u_2.

4.6 A proposed model for a managed multispecies fishery (May et al., 1979) is given by

$$\dot{x}_1 = x_1(1 - x_1 - x_2 - u_1)$$

$$\dot{x}_2 = \beta x_2\left(1 - \gamma\frac{x_2}{x_1} - u_2\right)$$

where the dot denotes differentiation with respect to time, x_1 = biomass of krill, x_2 = number of Baleen whales, u_1 = harvesting effort by krill fishery, u_2 = harvesting effort of whale fishery, β = ratio of intrinsic growth rates (a constant), and γ = ratio of several other constants associated with the model. What is the admissible range of constant harvesting efforts that will yield nonnegative equilibrium solutions for x_1 and x_2? An equilibrium solution corresponds to $\dot{x}_1 = \dot{x}_2 = 0$. Draw a sketch of the set of all possible equilibrium solutions in state space. For all admissible whale fishing efforts u_2 determine the state space equilibrium values that correspond to a maximum yield for the krill (i.e., given a constant admissible u_2, determine u_1 to maximize $u_1 x_1$; then plot the corresponding equilibrium in state space for all admissible u_2). Similarly, given an admissible krill fishing effort u_1, determine the state space equilibrium values that correspond to a maximum yield, $\beta x_2 u_2$, for the whales.

4.7 Determine the global maximum for $H(x, u) = x^4 + u^4$ subject to $g(x, u) = (u-1)^2 + x^2 - 1 = 0$ directly using geometrical considerations. Now attempt to apply Corollary 4.1 to obtain the same solution. Try using the method of direct substitution to obtain a solution.

4.8 Use Theorem 2.3 to obtain the necessary conditions of Theorem 4.1.

4.9 The imposition of equality constraints may induce implicit constraints on the controls. Compare solutions to the following two problems.

(a) Minimize $G = (u-2)^2$ subject to $-1 \leqq u \leqq 1$.

(b) Minimize $G = (u-2)^2$ subject to $u - \sin x = 0$.

Chapter Five

CONTINUOUS STATIC GAMES

5.1 INTRODUCTION

Thus far we have considered parametric systems having one or more cost criteria. The discussion has been limited, however, to the case of a single decision maker (controller) who has sole control over the selection of all of the system parameters $y \in Y$, subject to specified equality and inequality constraints. In this chapter we consider the more general case of multiple decision makers, each with their own cost criterion. This generalization introduces the possibility of competition among the system controllers, called "players," and the optimization problem under consideration is therefore termed a "game" (von Neumann and Morgenstern, 1944). Each player in the game controls a specified subset of the system parameters (called his control vector) and seeks to minimize his own scalar[†] cost criterion, subject to specified constraints.

Applications of game theory may be found in economics, engineering, biology, and in many other areas. Competition among firms seeking to maximize their own profits, and competition for food and territory among biological species are but two examples. Pursuit-evasion games, such as missile-aircraft intercept problems, are examples both of military applications and of games involving dynamic systems.

Although game theory is uniquely suited for the analysis of multiple-controller competitive systems, it may also be employed effectively in the analysis of uncertain systems via a worst case analysis.

Three major classes of games are matrix games, continuous static games, and differential games. Matrix games (von Neumann and Morgenstern, 1944) derive their name from a discrete relationship between a finite (or

[†] More generally, a player's cost could be a vector. If so, we assume scalarization has been employed to produce a corresponding scalar cost (see Chapter 3).

countable) number of possible decisions and the corresponding costs. This relationship is conveniently represented in terms of a matrix (for two-player games) in which one player's decision corresponds to the selection of a row and the other player's decision corresponds to the selection of a column, with the corresponding entry (or entries) denoting the costs. In continuous static games the decision possibilities need not be discrete, for example, $u \in [0,1]$, and the decisions and costs are related in a continuous rather than a discrete manner. The game is static in the sense that no time history is involved in the relationship between costs and decisions. Differential games (Blaquière et al., 1969; Isaacs, 1965) are characterized by continuously varying costs along with a dynamical system governed by ordinary differential equations.

In this chapter we restrict our discussion to continuous static games. However, the concepts of optimality presented here are applicable to other games.

In the following games control notation is used and each player, $i=1,\dots,r$, selects his control vector† $u^i \in E^{s_i}$ seeking to minimize a scalar-valued criterion

$$G_i(x,u) \tag{5.1}$$

subject to n equality constraints

$$g(x,u)=0 \tag{5.2}$$

where $x \in E^n$ is the state and $u=(u^1,\dots,u^r) \in E^s$, $s=s_1+\cdots+s_r$, is the *composite control*. The composite control is required to be an element of a regular control constraint set $\Omega \subseteq E^s$ of the form

$$\Omega=\{u \in E^s \mid h(x,u) \geqq 0\} \tag{5.3}$$

where $x=\xi(u)$ is the solution to (5.2) given u. The functions $G_i(\cdot): E^n \times E^s \to E^1$, $g(\cdot): E^n \times E^s \to E^n$, and $h(\cdot): E^n \times E^s \to E^q$ are assumed to be C^1, with

$$\left| \frac{\partial g(x,u)}{\partial x} \right| \neq 0, \tag{5.4}$$

in a ball about a solution point (x,u).

Note that the control constraints defined by (5.3) may be coupled through the state and each of the players' controls. As noted in Chapter 4, if

†Supercripts denote vectors, subscripts denote scalars, components of vectors, and sets.

possible, control and state variables should be selected so that $h(\cdot)$ is not a function of x. In many game situations it is also desirable to have $h(\cdot)$ uncoupled in each of the players' controls. For many problems complete decoupling is possible, that is, the inequality constraints may be written in the form $h^i(u^i) \geqq 0$. The generality of (5.3) is included to allow for flexibility in formulating problems and for compactness in stating necessary conditions of optimality.

In the following sections we frequently denote the control for a particular player i by u^i and use v to denote the composite control of the remaining players. In two-player games we often use u and v for the controls of Players 1 and 2, respectively.

As in the case of vector minimization, a concept for a game theoretic solution must be specified. Several solution concepts are possible, including the Pareto-minimal concept discussed in Chapter 3. In order to arrive at a choice for his control vector, a given player may examine a number of solution concepts. How he uses these concepts depends not only on information concerning the nature of the other players, but on his own personality as well. A given player may or may not play rationally, may or may not lie, cheat, cooperate, bargain, and so on. All of these factors must be considered by a player in making the ultimate choice of his control. In this chapter we examine some solution concepts that provide a given player with information that he may use, in conjunction perhaps with other information concerning the other players, to make his control choice.

In addition to the Pareto-minimal (cooperative) concept, we will examine the Nash equilibrium concept, the min-max (security) concept, and the Stackelberg leader-follower concept. Before introducing these concepts, however, we first develop the useful idea of a rational reaction set.

5.2 RATIONAL REACTION SETS

In determining his controls, each player must consider the possible choices of the other players. To this end we introduce the concept of rational reaction sets (Simaan and Cruz, 1973). Consider Player i with cost function $G_i(x, u^i, v)$ where Player i controls u^i, v is controlled by the other players, and $x = \xi(u^i, v)$ is the solution to (5.2).

DEFINITION 5.1 *The rational reaction set D_i for Player i* is defined by

$$D_i = \left\{ (u^i, v) \in \Omega \,|\, G_i\big(\xi(u^i, v), u^i, v\big) \leqq G_i\big(\xi(\bar{u}^i, v), \bar{u}^i, v\big) \right.$$
$$\left. \text{for all } \bar{u}^i \text{ such that } (\bar{u}^i, v) \in \Omega \right\}. \tag{5.5}$$

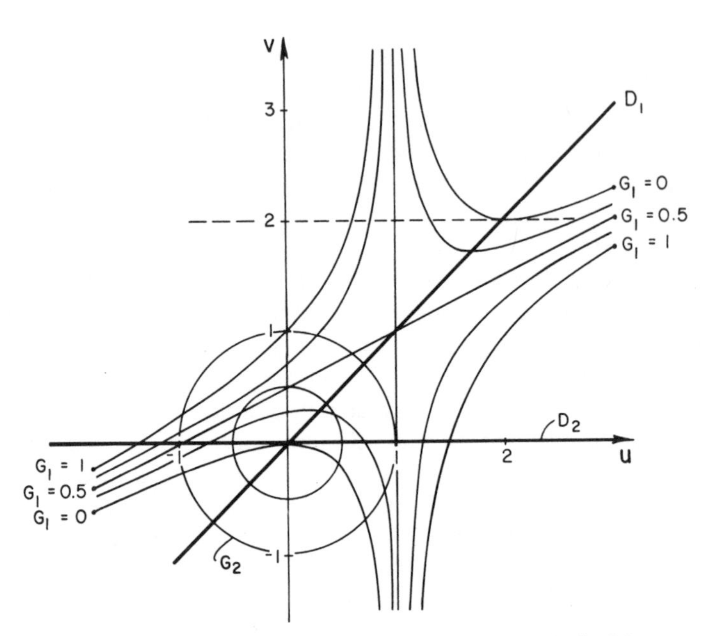

Figure 5.1. The rational reaction sets for Example 5.1.

Thus if the other players were to announce their composite control as v, Player i would rationally select u^i such that $(u^i, v) \in D_i$.

Example 5.1 In a game with two players, suppose Player 1 and Player 2, respectively, have the costs

$$G_1(\cdot) = -uv + \tfrac{1}{2}u^2 + v$$

$$G_2(\cdot) = u^2 + v^2$$

where Player 1 controls $u \in E^1$ and Player 2 controls $v \in E^1$. For a given v (e.g., $v=2$ in Figure 5.1) $G_1(\cdot)$ is minimized by selecting $u=v$. Thus the rational reaction set for Player 1 is

$$D_1 = \{(u,v) \in E^2 | u=v\}.$$

Similarly, the rational reaction set for Player 2 is

$$D_2 = \{(u,v) \in E^2 | v=0\}.$$

Figure 5.1 illustrates curves of constant $G_1(\cdot)$ and $G_2(\cdot)$ and the rational reaction sets for Players 1 and 2.

Notice in Definition 5.1 that $u=(u^i, v)\in D_i$ requires that u, along with the state $x=\xi(u)$ determined from (5.2), must satisfy the control constraints (5.3). In particular, a point (u^i, v) in the rational reaction set D_i for Player i is determined by selecting a composite control v for the other players and then choosing u^i to minimize $G_i(\cdot)$ subject to the constraints (5.2) and (5.3). Since these constraints are generally coupled in both the state x and the control $u=(u^i, v)$, there may be values of (u^i, v) not on the rational reaction set that violate the constraints (5.2) and (5.3). Clearly, in the *analysis* of a game such a point $u=(u^i, v)\notin\Omega$ is simply not an admissible control. A conceptual difficulty may arise however, associated with the actual *play* of the game. For example, players choosing their controls independently and then announcing their choice simultaneously may produce a control $u=(u^1,\dots, u^r)\notin\Omega$.

This difficulty associated with the play of the game under coupled constraints could be eliminated by some convenient device, such as a referee. In many applications the control constraints can be decoupled, that is, $h^i(u^i)\geqq 0$, which automatically eliminates the difficulty.

The rational reaction sets for the players in a game are used in the formal analysis of various solution concepts. As an introduction consider a two-player game in which Player 1 controls $u\in E^1$ and Player 2 controls $v\in E^1$ with $(u, v)\in\Omega$ and Ω defined by the region shown in Figure 5.2. Assume the cost for Player 1 is $G_1(u, v)$ and the cost for Player 2 is $G_2(u, v)$. Suppose that the rational reaction sets for each player have been determined using Definition 5.1 and are as shown in Figure 5.2. It is possible for the rational reaction sets to contain part of the boundary of Ω as indicated. Furthermore, a rational reaction set need not be a connected set and may have "corner" points.

The rational reaction set D_1 is obtained by assuming Player 2 announces his control v first. Assuming that Player 1 is "rational" and will indeed choose a control $u=u(v)$ in the rational reaction set D_1, then Player 2 can invoke a number of particular solution concepts that may be of interest. For example, Player 2 may be a good Samaritan and choose v to

$$\text{(min-min}-1) \qquad \text{minimize } G_1(u, v) \text{ subject to } (u, v)\in D_1,$$

which results in a particular *Pareto-minimal* (i.e., cooperative) solution. Player 2 may be diabolical and choose v to

$$\text{(min-max}-1) \qquad \text{maximize } G_1(u, v) \text{ subject to } (u, v)\in D_1,$$

which results in a *min-max* solution with respect to Player 1. Alternatively, Player 2, knowing $u(v)$, may look after his own interests and choose v to

$$(\text{S}_2) \qquad \text{minimize } G_2(u, v) \text{ subject to } (u, v)\in D_1,$$

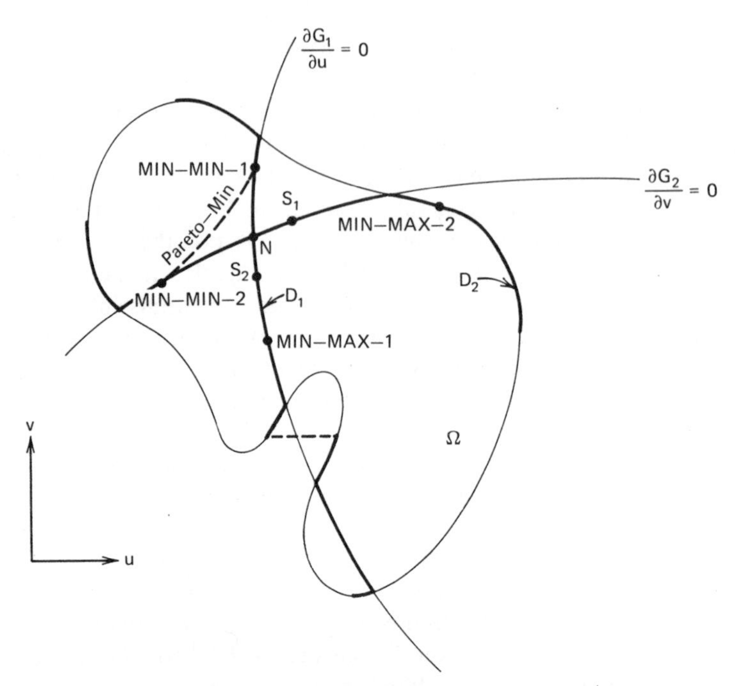

Figure 5.2. Rational reaction sets and game solutions.

which corresponds to a *Stackelberg* solution with Player 2 as Leader. Player 2 may also decide to play on his own rational reaction set, in which case he chooses v so that

(N) $$(u,v)\in D_1 \cap D_2,$$

which results in a *Nash* equilibrium solution. Clearly, numerous other solutions are possible. Indeed, every point on D_1 corresponds to some choice by Player 2.

In a similar vein, if Player 1 must announce his control u first and if Player 2 will indeed choose a control $v=v(u)$ in the rational reaction set D_2, then Player 1 can also examine a number of particular solutions. As a good Samaritan, Player 1 could choose u to

(min-min−2) minimize $G_2(u,v)$ subject to $(u,v)\in D_2$,

which results in a *Pareto-minimal* solution. As a diabolical individual, Player 1 could choose u to

(min-max−2) maximize $G_2(u,v)$ subject to $(u,v)\in D_2$,

which results in a *min-max* solution with respect to Player 2. Looking after his own interests, Player 1 could choose u to

$$(\mathrm{S}_1) \qquad \text{minimize } G_1(u,v) \text{ subject to } (u,v)\in D_2,$$

which corresponds to a *Stackelberg* solution with Player 1 as Leader. To play on his own rational reaction set, Player 1 would choose u so that

$$(\mathrm{N}) \qquad (u,v)\in D_1\cap D_2,$$

which results in a *Nash* equilibrium solution.

The various solution possibilities are indicated in Figure 5.2. Note that, in general, there are additional Pareto-minimal solutions that lie off of the two rational reaction sets. Cooperation between the players would be required in order to play these additional points, since they are not "rational" in the sense that they lie off of both rational reaction sets. However, under cooperation it would be "irrational" to play a non-Pareto-minimal point.

5.3 COST PERTURBATIONS AND TANGENT CONES

Before introducing formal definitions for the various solution concepts, it is convenient to first develop a general relation for the perturbations induced in a cost function $G_i(x,u)$ due to local perturbations about a nominal composite control u. This relation is used in the development of necessary conditions for all of the solution concepts.

At a point $u\in\Omega$ let $B\subset E^s$ denote a ball about u and let $x=\xi(u)$ denote the corresponding solution to $g(x,u)=0$. Let $T\subseteq E^s$ denote the tangent cone to Ω at u and let $\delta u(\cdot)$ generate a tangent vector $e\in T$, that is, let $u+\alpha\,\delta u(\alpha)\in\Omega$ for all sufficiently small $\alpha>0$ and $\delta u(\alpha)\to e$ as $\alpha\to 0$.

For each Player i, $i=1,\dots,r$, the first-order approximation theorem yields

$$\begin{aligned}\delta G_i &\triangleq G_i[\xi[u+\alpha\,\delta u(\alpha)],u+\alpha\,\delta u(\alpha)]-G_i[\xi(u),u]\\ &=\left[\frac{\partial G_i}{\partial x}\frac{\partial \xi}{\partial u}+\frac{\partial G_i}{\partial u}\right]\alpha e+\tilde{R}_i(\alpha)\end{aligned} \tag{5.6}$$

where $\tilde{R}_i(\alpha)/\alpha\to 0$ as $\alpha\to 0$ and all quantities are evaluated at (x,u). Applying the first-order approximation theorem to

$$g[\xi[u+\alpha\,\delta u(\alpha)],u+\alpha\,\delta u(\alpha)]=g[\xi(u),u]=0,$$

we have

$$0=\left[\frac{\partial g}{\partial x}\frac{\partial \xi}{\partial u}+\frac{\partial g}{\partial u}\right]\alpha e+\bar{R}(\alpha)$$

where $\bar{R}(\alpha)/\alpha \to 0$ as $\alpha \to 0$. Thus in view of (5.4) we have

$$\frac{\partial \xi}{\partial u}\alpha e=-\left[\frac{\partial g}{\partial x}\right]^{-1}\left[\frac{\partial g}{\partial u}\alpha e+\bar{R}(\alpha)\right] \tag{5.7}$$

for all $e \in T$.

Combining (5.6) and (5.7) yields

$$\delta G_i=\left[\frac{\partial G_i}{\partial u}-\frac{\partial G_i}{\partial x}\left[\frac{\partial g}{\partial x}\right]^{-1}\frac{\partial g}{\partial u}\right]\alpha e+R_i(\alpha) \tag{5.8}$$

for $e \in T$, where $R_i(\alpha)/\alpha \to 0$ as $\alpha \to 0$. Defining

$$\gamma^T(i) \triangleq \frac{\partial G_i}{\partial x}\left[\frac{\partial g}{\partial x}\right]^{-1} \tag{5.9}$$

and

$$J_i[x,u,\gamma(i)] \triangleq G_i(x,u)-\gamma^T(i)g(x,u), \tag{5.10}$$

we may write (5.8) as

$$\delta G_i=\frac{\partial J_i}{\partial u}\alpha e+R_i(\alpha) \tag{5.11}$$

for $e \in T$. Note from (5.9) and (5.10) that $\gamma(i)$ is the (unique) solution to

$$\frac{\partial J_i}{\partial x}=0. \tag{5.12}$$

For $u+\alpha\,\delta u(\alpha) \in \Omega$ for all sufficiently small $\alpha>0$ we have

$$h[\xi[u+\alpha\,\delta u(\alpha)],u+\alpha\,\delta u(\alpha)] \geqq 0.$$

Thus the first-order approximation theorem, along with (5.7) and the regularity assumption, yield the following result for T, the tangent cone to Ω

at u:

$$T=\left\{e\in E^{s}\left|\left[\frac{\partial\hat{h}}{\partial u}-\frac{\partial\hat{h}}{\partial x}\left[\frac{\partial g}{\partial x}\right]^{-1}\frac{\partial g}{\partial u}\right]e\geqq 0\right.\right\} \tag{5.13}$$

where $\hat{h}(\cdot)$ denotes the active inequality constraints at $u=(u^i, v)$.

In discussing a particular Player i with control $u^i \in E^{s_i}$, it is convenient to focus on the control constraints induced on Player i by a choice v of the composite control of the remaining players. Thus for specified v, with $u=(u^i, v)$, we have $u^i \in U_i$, where

$$U_i=\left\{u^i \in E^{s_i} \mid h\left[\xi(u^i, v), u^i, v\right]\geqq 0\right\}. \tag{5.14}$$

If $U_i \subseteq E^{s_i}$ is regular at u^i for a given v, then in a fashion analogous to the development of (5.13), the tangent cone $T_i \subseteq E^{s_i}$ to $U_i \in E^{s_i}$ at u^i is given by

$$T_i=\left\{e^i \in E^{s_i}\left|\left[\frac{\partial\hat{h}}{\partial u^i}-\frac{\partial\hat{h}}{\partial x}\left[\frac{\partial g}{\partial x}\right]^{-1}\frac{\partial g}{\partial u^i}\right]e^i\geqq 0\right.\right\} \tag{5.15}$$

where $\hat{h}(\cdot)$ denotes the active inequality constraints at $u=(u^i, v)$.

With coupled control constraints it is possible that some points $u^i \in U_i$ may not be regular points of U_i, even though (u^i, v) is a regular point of Ω. This may occur, for example, when the choice for v produces a set U_i that contains only a single point u_i. This situation is discussed in more detail in Example 5.15 and leads to the following definition.

DEFINITION 5.2 A point $u=(u^i, v)\in\Omega$, $i=1,\ldots, r$, is a *completely regular point* if and only if

(i) u is a regular point of Ω and
(ii) u^i is a regular point of U_i for each $i=1,\ldots, r$.

5.4 NASH EQUILIBRIUM SOLUTIONS

The Nash equilibrium solution concept (Nash, 1951; Starr and Ho, 1967, 1969) is tailored for the situation in which coalitions among players are not possible. It is assumed that the players act independently, without collaboration with any of the other players, and that each player seeks to minimize his own cost function. The information available to each player consists of the cost functions and constraints for each player, that is, equations (5.1)–(5.3).

DEFINITION 5.3 A point $\hat{u} \in \Omega$ is a *Nash equilibrium point* if and only if for each $i=1,\ldots,r$

$$G_i[\xi(\hat{u}), \hat{u}] \leq G_i[\xi(u^i, \hat{v}), u^i, \hat{v}] \tag{5.16}$$

for all $u^i \in U_i$ where $\hat{u}=(\hat{u}^i, \hat{v}) \in \Omega$, U_i is defined by (5.14), and $x=\xi(u)$ is the solution to (5.2). For a local Nash equilibrium point replace U_i by $B_i \cap U_i$ for some ball $B_i \subset E^{s_i}$ centered at $\hat{u}^i$.

From Definition 5.3 we see that, if the other players employ their Nash controls, then Player i cannot benefit by unilaterally changing his control from $\hat{u}^i$ to some other control $u^i \in U_i$. The assumption that the other players' controls remain fixed (at their Nash values $\hat{v}$) is a fundamental ingredient in the Nash equilibrium concept. If this assumption does not hold (i.e., if $v \neq \hat{v}$), then in general, in satisfying (5.16), $\hat{u}^i$ may not minimize $G_i[\xi(u^i, v), u^i, v]$. For example, in the actual play of a game a coalition of two or more players might form in which one player, to the detriment of Player i, agrees to play a non-Nash control in order to produce an overall benefit to the coalition.

It follows from Definition 5.3 that any Nash point $(\hat{u}^i, \hat{v})$ must belong to D_i, the rational reaction set for Player i, for each $i=1,\ldots,r$.

Consider a two-player game with scalar controls $u \in E^1$ and $v \in E^1$, and cost functions $G_1(u, v)$ and $G_2(u, v)$ for Players 1 and 2, respectively, where the lines of constant cost and the rational reaction sets for Players 1 and 2 are as depicted in Figure 5.3. If Player 1 is rational, then for any control v announced by Player 2, Player 1 will pick u such that the corresponding

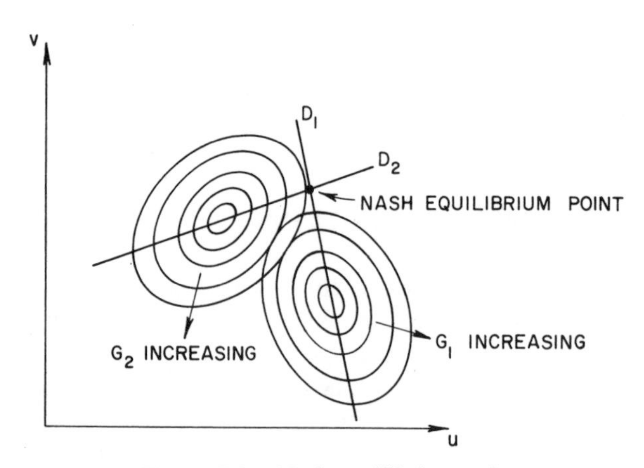

Figure 5.3. Nash equilibrium point.

point (u, v) is on the rational reaction set D_1. Similarly, if rational, Player 2 would respond to an announced Player 1 control u by picking v such that (u, v) lies on the rational reaction set D_2. The Nash equilibrium point lies in $D_1 \cap D_2$. If the sets D_1 and D_2 did not intersect, then a Nash equilibrium solution would not exist.

For the general case of r players we have the following result for a Nash equilibrium point.

Lemma 5.1 A point $\hat{u}=(\hat{u}^1,\ldots,\hat{u}^r)\in\Omega$ is a Nash equilibrium point if and only if

$$\hat{u}\in\bigcap_{i=1}^{r} D_i. \tag{5.17}$$

Proof The lemma follows directly from Definition 5.1 and Definition 5.3. ■

From Lemma 5.1 if

$$\bigcap_{i=1}^{r} D_i=\varnothing, \tag{5.18}$$

then a Nash equilibrium solution does not exist.

Example 5.2 Consider the following "two-player zero-sum" game, that is, $G_1(\cdot)=-G_2(\cdot)$, defined by

$$G_1(\cdot)=uv$$

$$G_2(\cdot)=-uv$$

where Player 1 chooses $u\in E^1$ and Player 2 chooses $v\in E^1$ with $-1\leqq u\leqq 2$ and $-1\leqq v\leqq 2$. For $v>0$, $G_1(\cdot)$ is minimized at $u=-1$. For $v<0$, $u=2$. For $v=0$, $u\in[-1,2]$. Thus the rational reaction set for Player 1 is

$$D_1=\{(u,v)\in E^2 \mid u=\bar{u}(v)\}$$

where

$$\bar{u}(v)=\begin{cases} -1 & \text{if} \quad v>0 \\ 2 & \text{if} \quad v<0 \\ u\in[-1,2] & \text{if} \quad v=0. \end{cases}$$

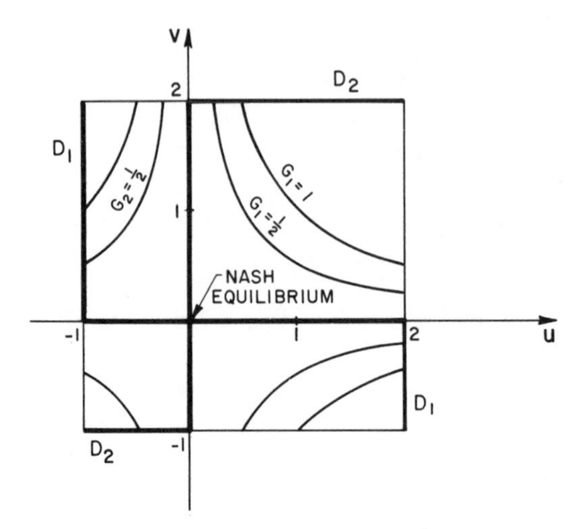

Figure 5.4. Nash equilibrium point for Example 5.2.

In a similar manner the rational reaction set for Player 2 is

$$D_2 = \{(u,v) \in E^2 | v = \bar{v}(u)\}$$

where

$$\bar{v}(u) = \begin{cases} -1 & \text{if} \quad u<0 \\ 2 & \text{if} \quad u>0 \\ v \in [-1,2] & \text{if} \quad u=0, \end{cases}$$

as illustrated in Figure 5.4. From Lemma 5.1 the Nash equilibrium points $(\hat{u}, \hat{v})$ must lie in the set

$$D_1 \cap D_2 = \{(0,0)\}.$$

Thus $(\hat{u}, \hat{v}) = (0,0)$ is the unique Nash equilibrium point. Notice that at the Nash equilibrium point neither player can benefit by a unilateral change in his control. But suppose it was guaranteed that one player would deviate from his Nash control. Then, in this example, the other player should also deviate. Note also that, if v were constrained to $0 \leqq v \leqq 2$, then every point in the set $\{(u,v) \in E^2 | v=0, \ u \in [-1,0]\}$ would be a Nash equilibrium point.

The procedure used in Example 5.2 was based solely on the definition of a Nash equilibrium point. We now develop first-order necessary conditions for local Nash equilibrium points in terms of the tangent cones to the control constraint sets U_i, induced on Player i by the composite control v chosen by the other players.

Let $T \subseteq E^s$ be the tangent cone to Ω at

$$u = \begin{bmatrix} u^1 \\ \vdots \\ u^r \end{bmatrix} \tag{5.19}$$

and let $T_i \subseteq E^{s_i}$ denote the tangent cone to U_i, where U_i and T_i are defined by (5.14) and (5.15), respectively. Note that if u is a completely regular point then $u + \alpha\,\delta u(\alpha) \in \Omega$ with $\delta u(\alpha) \to e \in T$ as $\alpha \to 0$ implies $u^i + \alpha\,\delta u^i(\alpha) \in U_i$ with $\delta u^i(\alpha) \to e^i \in T_i$ as $\alpha \to 0$ where

$$\delta u(\cdot) = \begin{bmatrix} \delta u^1(\cdot) \\ \vdots \\ \delta u^r(\cdot) \end{bmatrix} \tag{5.20}$$

and

$$e = \begin{bmatrix} e^1 \\ \vdots \\ e^r \end{bmatrix}. \tag{5.21}$$

In terms of the tangent cones T_i we establish the following principal result for a local Nash equilibrium point.

Lemma 5.2 If $\hat{u} \in \Omega$ is a local Nash equilibrium point for the game (5.1)–(5.3) and if $\hat{x} = \xi(\hat{u})$ is the solution to $g(x, \hat{u}) = 0$, then for each $i = 1, \ldots, r$ there exists a vector $\gamma(i) \in E^n$, defined by

$$\frac{\partial J_i[\hat{x}, \hat{u}, \gamma(i)]}{\partial x} = 0, \tag{5.22}$$

such that[†]

$$\frac{\partial J_i[\hat{x}, \hat{u}, \gamma(i)]}{\partial u^i} e^i \geqq 0 \tag{5.23}$$

[†] Repeated indices do not imply summation, although (5.23) is a "dot product" expression.

for all $e^i \in T_i$, where

$$J_i[x, u, \gamma(i)] \triangleq G_i(x, u) - \gamma^T(i) g(x, u). \tag{5.24}$$

Proof From (5.11) and Definition 5.3, for each $i = 1, \ldots, r$, at a Nash equilibrium point we have

$$\delta G_i = \frac{\partial J_i}{\partial u} \alpha e + R_i(\alpha) \geqq 0 \tag{5.25}$$

for all $e \in T$, where e is of the form (5.21) with

$$e^j = 0 \qquad \forall j \in \{1, \ldots, r\}, j \neq i. \tag{5.26}$$

Combining (5.25) and (5.26) yields

$$\frac{\partial J_i}{\partial u^i} \alpha e^i + R_i(\alpha) \geqq 0$$

for all $e^i \in T_i$. Dividing by $\alpha > 0$ and taking the limit as $\alpha \to 0$ establishes the lemma. ■

Necessary conditions usable for determining Nash equilibrium points may now be stated in the following theorem.

Theorem 5.1 If $\hat{u} \in \Omega$ is a completely regular local Nash equilibrium point for the game (5.1)–(5.3) and $\hat{x} = \xi(\hat{u})$ is the solution to $g(x, \hat{u}) = 0$, then for each $i = 1, \ldots, r$ there exists a vector $\lambda(i) \in E^n$ and a vector $\mu(i) \in E^q$ such that

$$\frac{\partial L_i[\hat{x}, \hat{u}, \lambda(i), \mu(i)]}{\partial x} = 0, \tag{5.27}$$

$$\frac{\partial L_i[\hat{x}, \hat{u}, \lambda(i), \mu(i)]}{\partial u^i} = 0 \tag{5.28}$$

$$g(\hat{x}, \hat{u}) = 0 \tag{5.29}$$

$$\mu^T(i) h(\hat{x}, \hat{u}) = 0 \tag{5.30}$$

$$h(\hat{x}, \hat{u}) \geqq 0 \tag{5.31}$$

$$\mu(i) \geqq 0 \tag{5.32}$$

where

$$L_i[x, u, \lambda(i), \mu(i)] \triangleq G_i(x,u) - \lambda^T(i)g(x,u) - \mu^T(i)h(x,u). \quad (5.33)$$

Proof For each $i = 1, \ldots, r$ consider the cone

$$K_i = \left\{ y \in E^{s_i} \,|\, y^T = \mu^T(i)\left[\frac{\partial h}{\partial u^i} - \frac{\partial h}{\partial x}\left[\frac{\partial g}{\partial x}\right]^{-1}\frac{\partial g}{\partial u^i}\right], \mu^T(i)h = 0, \mu(i) \geqq 0 \right\} \quad (5.34)$$

and its polar

$$K_i^* = \{ z \in E^{s_i} \,|\, y^T z \geqq 0 \ \forall\, y \in K_i \},$$

where all quantities are evaluated at $(\hat{x}, \hat{u})$. By the same procedure as in the proof of Lemma 2.5, since $\hat{u}^i$ is a regular point of U_i, the tangent cone T_i to U_i is given by

$$T_i = K_i^*.$$

From this result and Lemma 5.2 we have

$$\frac{\partial J_i[\hat{x}, \hat{u}, \gamma(i)]}{\partial u^i} z \geqq 0$$

for all $z \in K_i^*$ where $J_i(\cdot)$ is defined by (5.24) and $\gamma(i) \in E^n$ is defined by (5.22). Thus from Farkas' lemma

$$\frac{\partial J_i[\hat{x}, \hat{u}, \gamma(i)]}{\partial u^i} \in K_i.$$

Thus from (5.34) we have

$$\frac{\partial J_i}{\partial u^i} = \mu^T(i)\left[\frac{\partial h}{\partial u^i} - \frac{\partial h}{\partial x}\left[\frac{\partial g}{\partial x}\right]^{-1}\frac{\partial g}{\partial u^i}\right] \quad (5.35)$$

where $\mu(i)$ satisfies (5.30)–(5.32). Define

$$\lambda^T(i) = \gamma^T(i) - \mu^T(i)\frac{\partial h}{\partial x}\left[\frac{\partial g}{\partial x}\right]^{-1}, \quad (5.36)$$

where $\gamma(i)$ is defined by (5.10) and (5.12). Combining (5.12) and (5.36) yields

$$0=\frac{\partial J_i}{\partial x}=\frac{\partial G_i}{\partial x}-\left[\lambda^T(i)+\mu^T(i)\frac{\partial h}{\partial x}\left[\frac{\partial g}{\partial x}\right]^{-1}\right]\frac{\partial g}{\partial x},$$

which is equivalent to (5.27). Combining (5.10), (5.35) and (5.36) yields (5.28), which establishes the theorem. ■

Note that for the case of state-independent control constraints, that is, for $h(u)\geqq 0$, the vectors $\lambda(i)$ and $\gamma(i)$ are identical. Moreover, $\lambda(i)$ is then uniquely determined as a function of $\hat{x}$ and $\hat{u}$ from (5.27) since $\partial L_i/\partial x$ is independent of $\mu(i)$ for state-independent control constraints.

Example 5.3 Consider the two-player zero-sum game with

$$G_1(\cdot)=u^2-x-2v^2$$

$$G_2(\cdot)=-u^2+x+2v^2$$

$$g(\cdot)=x-u-v=0$$

where Player 1 selects $u\in E^1$ to minimize $G_1(\cdot)$ and Player 2 selects $v\in E^1$ to minimize $G_2(\cdot)$. Defining

$$L_1=u^2-x-2v^2-\lambda(1)(x-u-v)$$

$$L_2=-u^2+x+2v^2-\lambda(2)(x-u-v),$$

we have

$$\frac{\partial L_1}{\partial x}=0=-1-\lambda(1)$$

$$\frac{\partial L_2}{\partial x}=0=1-\lambda(2)$$

so that

$$\lambda(1)=-1$$

$$\lambda(2)=1.$$

Then

$$\frac{\partial L_1}{\partial u}=0=2u+\lambda(1)$$

$$\frac{\partial L_2}{\partial v}=0=4v+\lambda(2)$$

imply

$$\hat{u}=-\frac{\lambda(1)}{2}=\tfrac{1}{2}$$

$$\hat{v}=-\frac{\lambda(2)}{4}=-\tfrac{1}{4}.$$

Example 5.4 Two firms sell substitutable products and seek to maximize their profits through advertising. Using the model from Leitmann and Schmitendorf (1978), the equilibrium (steady-state) fractions of the market x_1 and x_2 that each firm receives are given by

$$0=g_1(\cdot)=-3x_1+u-u^2-x_1v \tag{5.37}$$

$$0=g_2(\cdot)=-2x_2+v-v^2-2x_2u \tag{5.38}$$

where $u \geqq 0$ and $v \geqq 0$ are the advertising expenditure rates for Firms 1 and 2, respectively, and $x_1 \geqq 0$, $x_2 \geqq 0$ with $x_1+x_2 \leqq 1$. Thus we have the inequality constraints

$$h_1(\cdot)=x_1 \geqq 0 \tag{5.39}$$

$$h_2(\cdot)=x_2 \geqq 0 \tag{5.40}$$

$$h_3(\cdot)=1-x_1-x_2 \geqq 0 \tag{5.41}$$

$$h_4(\cdot)=u \geqq 0 \tag{5.42}$$

$$h_5(\cdot)=v \geqq 0. \tag{5.43}$$

The steady-state profits of Firms 1 and 2 are taken, respectively, as

$$H_1(\cdot)=5x_1-u \tag{5.44}$$

$$H_2(\cdot)=3x_2-v. \tag{5.45}$$

Thus Firms 1 and 2 seek to minimize

$$G_1(\cdot) = -5x_1 + u \tag{5.46}$$

$$G_2(\cdot) = -3x_2 + v, \tag{5.47}$$

respectively.

Using Theorem 5.1, we define

$$\begin{aligned} L_1 = &-5x_1 + u - \lambda_1(1)(-3x_1 + u - u^2 - x_1 v) \\ &- \lambda_2(1)(-2x_2 + v - v^2 - 2x_2 u) - \mu_1(1)x_1 - \mu_2(1)x_2 \\ &- \mu_3(1)(1 - x_1 - x_2) - \mu_4(1)u - \mu_5(1)v \\ L_2 = &-3x_2 + v - \lambda_1(2)(-3x_1 + u - u^2 - x_1 v) \\ &- \lambda_2(2)(-2x_2 + v - v^2 - 2x_2 u) - \mu_1(2)x_1 - \mu_2(2)x_2 \\ &- \mu_3(2)(1 - x_1 - x_2) - \mu_4(2)u - \mu_5(2)v. \end{aligned}$$

The necessary conditions (5.27) and (5.28) for Nash equilibra are

$$\frac{\partial L_1}{\partial x_1} = 0 = -5 + \lambda_1(1)(3 + v) - \mu_1(1) + \mu_3(1) \tag{5.48}$$

$$\frac{\partial L_1}{\partial x_2} = 0 = 2\lambda_2(1)(1 + u) - \mu_2(1) + \mu_3(1) \tag{5.49}$$

$$\frac{\partial L_2}{\partial x_1} = 0 = \lambda_1(2)(3 + v) - \mu_1(2) + \mu_3(2) \tag{5.50}$$

$$\frac{\partial L_2}{\partial x_2} = 0 = -3 + 2\lambda_2(2)(1 + u) - \mu_2(2) + \mu_3(2) \tag{5.51}$$

$$\frac{\partial L_1}{\partial u} = 0 = 1 - \lambda_1(1)(1 - 2u) + 2\lambda_2(1)x_2 - \mu_4(1) \tag{5.52}$$

$$\frac{\partial L_2}{\partial v} = 0 = 1 + \lambda_1(2)x_1 - \lambda_2(2)(1 - 2v) - \mu_5(2) \tag{5.53}$$

where $\mu_j(i) \geqq 0$, $\mu_j(i)h_j = 0$, $i = 1, 2$, $j = 1, \ldots, 5$. From (5.48)–(5.51) we have

$$\lambda_1(1)=\frac{5+\mu_1(1)-\mu_3(1)}{3+v}$$

$$\lambda_2(1)=\frac{\mu_2(1)-\mu_3(1)}{2(1+u)}$$

$$\lambda_1(2)=\frac{\mu_1(2)-\mu_3(2)}{3+v}$$

$$\lambda_2(2)=\frac{3+\mu_2(2)-\mu_3(2)}{2(1+u)}.$$

Substituting these results into (5.52) and (5.53) yields

$$0=1-\frac{1-2u}{3+v}[5+\mu_1(1)-\mu_3(1)]+\frac{x_2}{1+u}[\mu_2(1)-\mu_3(1)]-\mu_4(1) \tag{5.54}$$

$$0=1+\frac{x_1}{3+v}[\mu_1(2)-\mu_3(2)]-\frac{1-2v}{2(1+u)}[3+\mu_2(2)-\mu_3(2)]-\mu_5(2). \tag{5.55}$$

From (5.37) and (5.38)

$$x_1=\frac{u(1-u)}{3+v} \tag{5.56}$$

$$x_2=\frac{v(1-v)}{2(1+u)}. \tag{5.57}$$

For x_1, x_2, u, and v nonnegative (5.56) and (5.57) imply $0 \leqq u \leqq 1$ and $0 \leqq v \leqq 1$. Using (5.56) and (5.57) and examining the problem of maximizing x_1+x_2, we find $\max(x_1+x_2)<1$. Hence $\mu_3(1)=\mu_3(2)=0$. Furthermore, $\mu_2(1)x_2=0$ and $\mu_1(2)x_1=0$. Thus (5.54) and (5.55) reduce to

$$0=1-\frac{(1-2u)[5+\mu_1(1)]}{3+v}-\mu_4(1) \tag{5.58}$$

$$0=1-\frac{(1-2v)[3+\mu_2(2)]}{2(1+u)}-\mu_5(2). \tag{5.59}$$

To examine boundary solutions suppose $u=0$. Then (5.58) cannot be satisfied with $0 \leqq v \leqq 1$, $\mu_1(1) \geqq 0$, and $\mu_4(1) \geqq 0$. Hence $u>0$, which implies $\mu_4(1)=0$. Alternately, suppose $u=1$. Then (5.58) cannot be satisfied with $0 \leqq v \leqq 1$, $\mu_4(1)=0$, and $\mu_1(1) \geqq 0$. Hence $0<u<1$, $x_1>0$, $\mu_1(1)=\mu_1(2)=\mu_4(1)=\mu_4(2)=0$, and (5.58) becomes

$$0=1-\frac{5(1-2u)}{3+v}. \tag{5.60}$$

Similarly, suppose $v=0$. Then (5.60) implies $u=0.2$ and (5.59) cannot be satisfied with $\mu_2(2)$ and $\mu_5(2)$ nonnegative. Hence $v>0$ and $\mu_5(2)=0$. If $v=1$, then (5.59) cannot be satisfied with $\mu_5(2)=0$, $u \geqq 0$, and $\mu_2(2) \geqq 0$. Hence $0<v<1$, $x_2>0$, $\mu_2(1)=\mu_2(2)=\mu_5(1)=\mu_5(2)=0$, and (5.59) becomes

$$0=1-\frac{3(1-2v)}{2(1+u)}. \tag{5.61}$$

Thus we have an interior solution with $\mu_j(i)=0$, $i=1,2$, $j=1,\ldots,5$, and (5.60) and (5.61) yield

$$\hat{u}=\tfrac{11}{58}$$

$$\hat{v}=\tfrac{3}{29}.$$

From (5.56) and (5.57) we have

$$\hat{x}_1=\frac{u(1-u)}{3+v}=0.0495$$

$$\hat{x}_2=\frac{v(1-v)}{2(1+u)}=0.0390.$$

The corresponding profits for each firm are

$$H_1=0.0580$$

$$H_2=0.0135.$$

Using (5.56) and (5.57) and substituting into (5.46) and (5.47) yields

$$G_1(u,v)=\frac{5u^2-2u+uv}{3+v}$$

$$G_2(u,v)=\frac{3v^2-v+2uv}{2(1+u)}$$

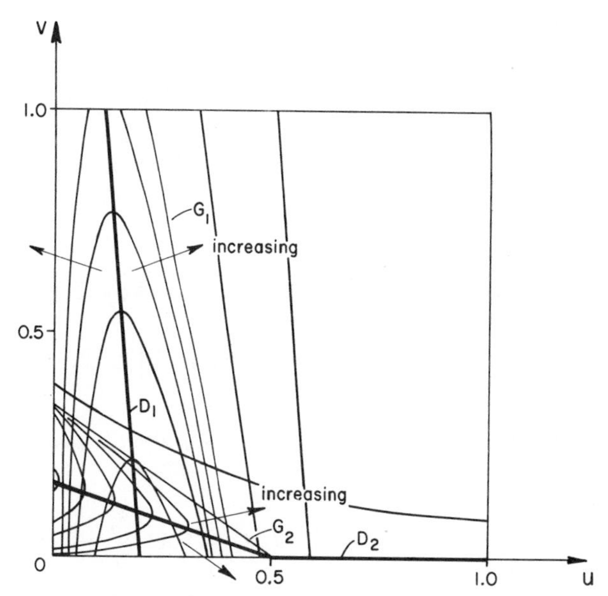

Figure 5.5. Geometry for Example 5.4.

where $0 \leqq u \leqq 1$ and $0 \leqq v \leqq 1$ account for all of the inequality constraints (5.39)–(5.43) since $h_3(\cdot)$ cannot be active. Lines of constant $G_1(\cdot)$ and $G_2(\cdot)$ and the rational reaction sets for the two firms are shown in Figure 5.5. The Nash equilibrium point is the point of intersection of the two sets.

5.5 MIN-MAX SOLUTIONS

In the Nash equilibrium solution each player tacitly assumes that all other players employ their Nash equilibrium controls. If the players, other than Player i, chose controls v that are not their Nash equilibrium controls, then the Nash equilibrium control for Player i may not minimize $G_i[\xi(u^i, v), u^i, v]$.

Player i may not feel secure in a belief that all other players will employ their Nash equilibrium controls. Instead of acting independently, in the pursuit of their own interests, one or more players may choose controls that tend to maximize the cost of Player i.

Under the min-max (security) solution concept (von Neumann and Morgenstern, 1944), Player i chooses his control under the assumption that all of the players have formed a coalition to maximize his cost. This approach is pessimistic in that it yields a worst case[†] result for Player i.

Since we concentrate on Player i, let his control be $u^i \in E^{s_i}$ and let $v \in E^{s-s_i}$ denote the controls for the other players. Let $G_i(x, u^i, v)$ denote the cost for player i and let $x = \xi(u^i, v)$ denote the solution to

$$g(x, u^i, v) = 0, \tag{5.62}$$

given controls u^i and v, where $(u^i, v) \in \Omega$ as defined by (5.3), that is,

$$\Omega = \left\{ (u^i, v) \in E^s \,|\, h\left[\xi(u^i, v), u^i, v \right] \geqq 0 \right\}. \tag{5.63}$$

DEFINITION 5.4 A point $u^* = (u^{i^*}, v^*) \in \Omega$ is a *min-max point for Player i* if and only if

$$G_i\left(\xi(u^{i^*}, v), u^{i^*}, v \right) \leqq G_i\left(\xi(u^*), u^* \right) \leqq G_i\left(\xi(u^i, v^*), u^i, v^* \right) \tag{5.64}$$

for all u^i, v such that $(u^{i^*}, v) \in \Omega$ and $(u^i, v^*) \in \Omega$. A point $(u^{i^*}, v^*) \in \Omega$ is a *local min-max point for Player i* if and only if (5.64) holds with Ω replaced by $B \cap \Omega$ for some ball $B \subset E^s$ centered at u^*.

To examine a min-max solution for Player i we employ D_i, the rational reaction set for Player i.

Lemma 5.3 A point $u^* = (u^{i^*}, v^*) \in \Omega$ is a min-max point for Player i if and only if $(u^{i^*}, v^*) \in D_i$ and

$$G_i\left(\xi(u^*), u^* \right) \geqq G_i\left(\xi(u^{i^*}, v), u^{i^*}, v \right) \qquad \forall (u^{i^*}, v) \in \Omega.$$

Proof The Lemma follows immediately from Definition 5.1 and Definition 5.4. ■

Let M_i denote a maximizing set for the other players, with respect to $G_i(\cdot)$, that is,

$$M^i = \left\{ (u^i, v) \in \Omega \,|\, G_i\left(\xi(u^i, v), u^i, v \right) \geqq G_i\left(\xi(u^i, \bar{v}), u^i, \bar{v} \right) \forall (u^i, \bar{v}) \in \Omega \right\}. \tag{5.65}$$

[†]For the optimist, who might want to consider the case where all of the other players want to help him rather than oppose him, see Chapter 4.

From (5.65), Definition 5.1, and Definition 5.4, a min-max point (if it exists) must satisfy

$$u^* = (u^{i^*}, v^*) \in D_i \cap M_i. \tag{5.66}$$

Note that min-max solutions (and the associated costs) are not necessarily unique, since $D_i \cap M_i$ may contain more than one point. Also note that in a two-player zero-sum game, that is, when $G_2(\cdot) = -G_1(\cdot)$, the min-max solutions for either player (if they exist) are also Nash solutions.

Example 5.5: In a two-player game let Player 1 control $u \in E^1$ and let Player 2 control $v \in E^1$. Let

$$G_1(u,v) = u^2 - v^2,$$

which Player 1 seeks to minimize. Then

$$D_1 = \{(u,v) | u = 0\}$$

and

$$M_1 = \{(u,v) | v = 0\}.$$

Hence there is a single, isolated min-max point for Player 1, given by $(u^*, v^*) = (0,0)$, as illustrated in Figure 5.6.

Example 5.6 Consider, as in Example 5.1,

$$G_1(u,v) = -uv + \tfrac{1}{2}u^2 + v$$

where $u \in E^1$ and $v \in E^1$ are the controls for Player 1 and for the other

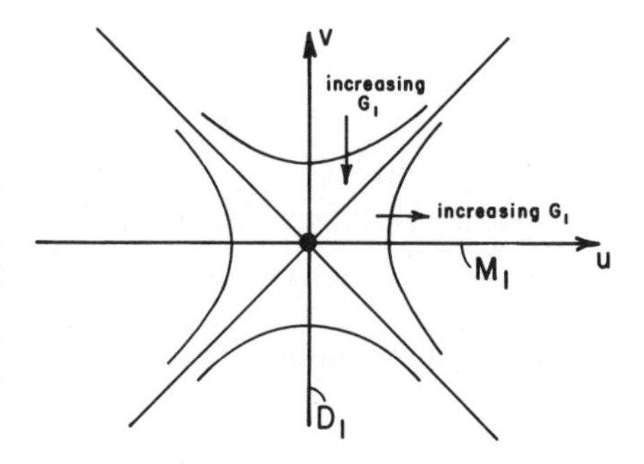

Figure 5.6. Min-max point geometry for Example 5.5.

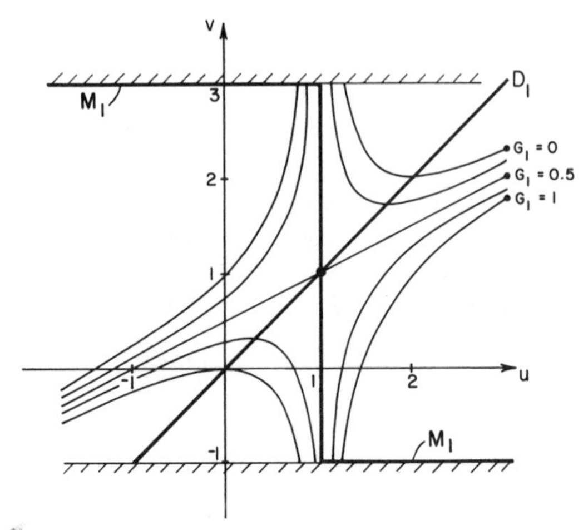

Figure 5.7. The sets D_1 and M_1 for Example 5.5 with $-1 \leqq v \leqq 3$.

players, respectively (see Figure 5.7). In the absence of constraints on v, a min-max solution for Player 1 does not exist, since for each $u \in E^1$

$$\sup_{v \in E^1} G_1(u, v) = +\infty.$$

If, however, v is constrained to $-1 \leqq v \leqq 3$, then the sets D_1 and M_1, illustrated in Figure 5.7, are

$$D_1 = \{(u, v) | u = v, \ -1 \leqq v \leqq 3\}$$

$$M_1 = \{(u, v) | v = \bar{v}(u)\}$$

where

$$\bar{v}(u) = \begin{cases} 3 & \text{if } u < 1 \\ -1 & \text{if } u > 1 \\ v \in [-1, 3] & \text{if } u = 1. \end{cases}$$

Since $D_1 \cap M_1 = \{(1, 1)\}$, the min-max point for Player 1 is $(u^*, v^*) = (1, 1)$.

As with the Nash equilibrium solutions, the min-max solution may not exist for a given problem.

Example 5.7 Consider a two-player game where Player 1 controls $u \in E^1$ and Player 2 controls $v \in E^1$ with $-1 \leqq v \leqq 3$. Player 1 seeks to minimize

$$G_1(\cdot) = (u - v)^2.$$

As illustrated in Figure 5.8, the sets D_1 and M_1 are

$$D_1 = \{(u, v) \in E^2 \,|\, u = v,\ -1 \leqq v \leqq 3\}$$

and

$$M_1 = \{(u, v) \in E^2 \,|\, v = \bar{v}(u)\}$$

where

$$\bar{v}(u) = \begin{cases} -1 & \text{if} \quad u > 1 \\ 3 & \text{if} \quad u < 1 \\ -1 \text{ or } 3 & \text{if} \quad u = 1. \end{cases}$$

Since $D_1 \cap M_1 = \varnothing$, a min-max solution does not exist for Player 1.

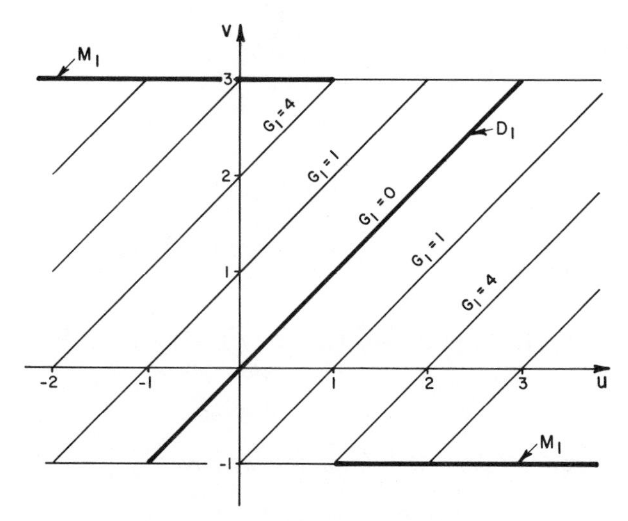

Figure 5.8. Sets D_1 and M_1 for Example 5.7 where a min-max solution does not exist.

Thus far we have only made use of the definition of a min-max solution for Player i. To develop necessary conditions for $u^*=(u^{i^*}, v^*)\in\Omega$ to be a min-max point for Player i, we employ the constraints $u^i\in U_i$ and $v\in V$ where the constraint sets U_i and V are defined at (u^{i^*}, v^*) by

$$U_i=\{u^i\in E^{s_i}|h[\xi(u^i, v^*), u^i, v^*]\geqq 0\} \tag{5.67}$$

$$V=\{v\in E^{s-s_i}|h[\xi(u^{i^*}, v), u^{i^*}, v]\geqq 0\}. \tag{5.68}$$

Let $T_i\subseteq E^{s_i}$ and $T_v\subseteq E^{s-s_i}$ denote the tangent cones to U_i and V, respectively, at $(u^{i^*}, v^*)\in\Omega$ and let $e^i\in T_i$ and $e^v\in T_v$ denote corresponding tangent vectors. We have the following fundamental necessary condition.

Lemma 5.4 If $u^*=(u^{i^*}, v^*)\in\Omega$ is a local min-max point for Player i for the game (5.1)–(5.3) and $x^*=\xi(u^*)$ is the solution to $g(x, u^*)=0$, then there exists a vector $\gamma(i)\in E^n$, defined by

$$\frac{\partial J_i[x^*, u^*, \gamma(i)]}{\partial x}=0, \tag{5.69}$$

such that

$$\frac{\partial J_i[x^*, u^*, \gamma(i)]}{\partial u^i}e^i\geqq 0 \tag{5.70}$$

for all $e^i\in T_i$ and

$$\frac{\partial J_i[x^*, u^*, \gamma(i)]}{\partial v}e^v\leqq 0 \tag{5.71}$$

for all $e^v\in T_v$, where

$$J_i[x, u, \gamma(i)]\triangleq G_i(x, u)-\gamma^T(i)g(x, u). \tag{5.72}$$

Proof From (5.11) and Definition 5.4, we have

$$0\leqq\frac{\partial J_i[x^*, u^*, \gamma(i)]}{\partial u^i}\alpha e^i+R_i(\alpha) \tag{5.73}$$

for all $e^i \in T_i$ and,

$$0 \geqq \frac{\partial J_i[x^*, u^*, \gamma(i)]}{\partial v} \alpha e^v + R_v(\alpha) \tag{5.74}$$

for all $e^v \in T_v$, where $R_i(\alpha)/\alpha \to 0$ and $R_v(\alpha)/\alpha \to 0$ as $\alpha \to 0$. Dividing (5.73) and (5.74) by $\alpha > 0$ and taking the limit as $\alpha \to 0$ establishes the lemma. ■

If u^* is a completely regular point of the constraint set Ω, then the tangent cones T_i and T_v are of the form (5.13), that is

$$T_i = \left\{ e^i \in E^{s_i} \middle| \left[\frac{\partial \hat{h}}{\partial u^i} - \frac{\partial \hat{h}}{\partial x} \left[\frac{\partial g}{\partial x} \right]^{-1} \frac{\partial g}{\partial u^i} \right] e^i \geqq 0 \right\} \tag{5.75}$$

$$T_v = \left\{ e^v \in E^{s-s_i} \middle| \left[\frac{\partial \hat{h}}{\partial v} - \frac{\partial \hat{h}}{\partial x} \left[\frac{\partial g}{\partial x} \right]^{-1} \frac{\partial g}{\partial v} \right] e^v \geqq 0 \right\} \tag{5.76}$$

where all quantities are evaluated at $(x^*, u^*) = (x^*, u^{i^*}, v^*)$ with $x^* = \xi(u^*)$ and where $\hat{h}(\cdot)$ denotes the active inequality constraints at (x^*, u^*).

Using (5.75) and (5.76) in Lemma 5.4, we have the following theorem for completely regular points.

Theorem 5.2 If $u^* = (u^{i^*}, v^*) \in \Omega$ is a completely regular local min-max point for Player i and if $x^* = \xi(u^*)$ is the solution to $g(x, u^*) = 0$, then there exist vectors $\lambda(i) \in E^n$, $\bar{\lambda}(i) \in E^n$, $\mu(i) \in E^q$, and $\bar{\mu}(i) \in E^q$ such that

$$\frac{\partial L_i[x^*, u^*, \lambda(i), \mu(i)]}{\partial x} = 0 \tag{5.77}$$

$$\frac{\partial L_i[x^*, u^*, \bar{\lambda}(i), \bar{\mu}(i)]}{\partial x} = 0 \tag{5.78}$$

$$\frac{\partial L_i[x^*, u^*, \lambda(i), \mu(i)]}{\partial u^i} = 0 \tag{5.79}$$

$$\frac{\partial L_i[x^*, u^*, \bar{\lambda}(i), \bar{\mu}(i)]}{\partial v} = 0 \tag{5.80}$$

$$g(x^*, u^*) = 0 \tag{5.81}$$

$$h(x^*, u^*) \geqq 0 \tag{5.82}$$

$$\mu^T(i)h(x^*, u^*)=0 \tag{5.83}$$

$$\bar{\mu}^T(i)h(x^*, u^*)=0 \tag{5.84}$$

$$\mu(i)\geqq 0 \tag{5.85}$$

$$\bar{\mu}(i)\leqq 0 \tag{5.86}$$

where

$$L_i[x, u, \lambda(i), \mu(i)]=G_i(x, u)-\lambda^T(i)g(x, u)-\mu^T(i)h(x, u) \tag{5.87}$$

and the partial derivatives of L_i are evaluated using the two sets of multipliers $\lambda(i)$, $\mu(i)$ and $\bar{\lambda}(i)$, $\bar{\mu}(i)$.

Proof Consider the cones

$$K_1=\left\{y(1)\in E^{s_i} \mid y^T(1)=\tilde{\mu}^T(1)\left[\frac{\partial h}{\partial u^i}-\frac{\partial h}{\partial x}\left[\frac{\partial g}{\partial x}\right]^{-1}\frac{\partial g}{\partial u^i}\right],\right.$$

$$\left.\tilde{\mu}^T(1)h(x^*, u^*)=0,\ \tilde{\mu}(1)\geqq 0\right\}$$

$$K_2=\left\{y(2)\in E^{s-s_i} \mid y^T(2)=\tilde{\mu}^T(2)\left[\frac{\partial h}{\partial v}-\frac{\partial h}{\partial x}\left[\frac{\partial g}{\partial x}\right]^{-1}\frac{\partial g}{\partial v}\right],\right.$$

$$\left.\tilde{\mu}^T(2)h(x^*, u^*)=0,\ \tilde{\mu}(2)\geqq 0\right\}$$

and their respective polars

$$K_1^*=\{z(1)\in E^{s_i} \mid y^T(1)z(1)\geqq 0\ \forall y(1)\in K_1\}$$

$$K_2^*=\{z(2)\in E^{s-s_i} \mid y^T(2)z(2)\geqq 0\ \forall y(2)\in K_2\}.$$

Then from Lemma 3.4 the tangent cones T_i and T_v, given by (5.56) and (5.57), satisfy

$$T_i=K_1^*$$

$$T_v=K_2^*.$$

From Farkas' lemma and Lemma 5.4

$$\frac{\partial J_i[x^*, u^*, \gamma(i)]}{\partial u^i} \in K_1 \tag{5.88}$$

$$-\frac{\partial J_i[x^*, u^*, \gamma(i)]}{\partial v} \in K_2 \tag{5.89}$$

where $\gamma(i)$ is defined by

$$\frac{\partial J_i[x^*, u^*, \gamma(i)]}{\partial x} = 0 \tag{5.90}$$

with $J_i(\cdot)$ defined by (5.72). Define

$$\lambda^T(i) = \gamma^T(i) - \tilde{\mu}^T(1)\frac{\partial h}{\partial x}\left[\frac{\partial g}{\partial x}\right]^{-1} \tag{5.91}$$

$$\bar{\lambda}^T(i) = \gamma^T(i) - \tilde{\mu}^T(2)\frac{\partial h}{\partial x}\left[\frac{\partial g}{\partial x}\right]^{-1} \tag{5.92}$$

With $L_i(\cdot)$ defined by (5.87) and choosing $\mu(i) = \tilde{\mu}(1) \geqq 0$, (5.77) follows from (5.90) by substituting for $\gamma(i)$ from (5.91). Similarly (5.79), (5.83), and (5.85) follow from (5.88), (5.91), and the definition of K_1. Choosing $\bar{\mu}(i) = -\tilde{\mu}(2) \leqq 0$ and substituting for $\gamma(i)$ from (5.92), (5.78) follows from (5.90). Similarly (5.80), (5.84), and (5.86) follow from (5.89) and the definition of K_2. ■

The necessary conditions of Theorem 5.2 simplify for state-independent inequality constraints, that is, $h(u^i, v) \geqq 0$. Then from (5.91) $\lambda(i) = \bar{\lambda}(i) = \gamma(i)$ and (5.77) and (5.78) become equivalent and independent of $\mu(i)$ and $\bar{\mu}(i)$. The necessary conditions simplify even further if the inequality constraints are decoupled, for example, if $h^1(u^i) \geqq 0$ and $h^2(v) \geqq 0$. In this case the components of $\mu(i)$ multiplying zero gradients ($\partial h^2/\partial u^i \equiv 0$) in (5.79) and the components of $\bar{\mu}(i)$ multiplying zero gradients ($\partial h^1/\partial v \equiv 0$) in (5.80) may be set equal to zero. Hence a single μ-vector, $\mu(i)$, may be employed, with the appropriate signs for the components of $\mu(i)$, that is, $\mu_j(i) \geqq 0$ ($\leqq 0$) if $h_j(\cdot) \geqq 0$ applies to the minimizing (maximizing) player.

Example 5.8 Consider the problem in Example 5.4 with equality constraints

$$0 = g_1(\cdot) = -3x_1 + u - u^2 - x_1 v \tag{5.93}$$

$$0 = g_2(\cdot) = -2x_2 + v - v^2 - 2x_2 u \tag{5.94}$$

and costs

$$G_1(\cdot) = -5x_1 + u \tag{5.95}$$

$$G_2(\cdot) = -3x_2 + v \tag{5.96}$$

where Player 1 controls u and Player 2 controls v with $0 \leqq u \leqq 1$ and $0 \leqq v \leqq 1$. The additional control constraints given here automatically satisfy the state inequality constraints of Example 5.4.

We determine the min-max solutions for each player. Since the inequality constraints are independent of x, we have a single λ-vector, denoted by $\lambda(i)$, for determining the min-max solution for Player i. Furthermore, since the inequality constraints are decoupled, we can employ the $L_i(\cdot)$ function for Player i using a single μ-vector, $\mu(i)$, with the appropriate sign conditions for the components of $\mu(i)$.

To determine the min-max solution for Player 1 define

$$\begin{aligned} L_1 = {} & -5x_1 + u - \lambda_1(1)(-3x_1 + u - u^2 - x_1 v) \\ & -\lambda_2(1)(-2x_2 + v - v^2 - 2x_2 u) - \mu_1(1)u - \mu_2(1)(1-u) \\ & -\mu_3(1)v - \mu_4(1)(1-v) \end{aligned}$$

where $\mu_1(1), \mu_2(1) \geqq 0$ and $\mu_3(1), \mu_4(1) \leqq 0$. Employing Theorem 5.2, we have

$$\frac{\partial L_1}{\partial x_1} = 0 = -5 + \lambda_1(1)(3 + v) \tag{5.97}$$

$$\frac{\partial L_1}{\partial x_2} = 0 = 2\lambda_2(1)(1 + u) \tag{5.98}$$

$$\frac{\partial L_1}{\partial u} = 0 = 1 - \lambda_1(1)(1 - 2u) + 2\lambda_2(1)x_2 - \mu_1(1) + \mu_2(1) \tag{5.99}$$

$$\frac{\partial L_1}{\partial v} = 0 = \lambda_1(1)x_1 - \lambda_2(1)(1 - 2v) - \mu_3(1) + \mu_4(1). \tag{5.100}$$

Since $u \geqq 0$ and $v \geqq 0$, (5.97) and (5.98) yield

$$\lambda_1(1) = \frac{5}{3+v}$$

$$\lambda_2(1) = 0$$

and (5.99) and (5.100) become

$$\frac{\partial L_1}{\partial u} = 0 = 1 - \frac{5(1-2u)}{3+v} - \mu_1(1) + \mu_2(1) \tag{5.101}$$

$$\frac{\partial L_1}{\partial v} = 0 = \frac{5x_1}{3+v} - \mu_3(1) + \mu_4(1). \tag{5.102}$$

Solving for x_1 from (5.93) yields

$$x_1 = \frac{u(1-u)}{3+v}. \tag{5.103}$$

Combining (5.102) and (5.103), we have

$$\frac{\partial L_1}{\partial v} = 0 = \frac{5u(1-u)}{(3+v)^2} - \mu_3(1) + \mu_4(1). \tag{5.104}$$

Suppose $0 < v^* < 1$. Then $\mu_3(1) = \mu_4(1) = 0$ and (5.104) yields $u = 0$ or $u = 1$. For $u = 0$, $\mu_2(1) = 0$ and (5.101) yields

$$\mu_1(1) = 1 - \frac{5}{3+v} \geqq 0,$$

which cannot be satisfied by $0 \leqq v \leqq 1$. For $u = 1$, $\mu_1(1) = 0$ and (5.101) yields

$$\mu_2(1) = -1 - \frac{5}{3+v} \geqq 0,$$

which cannot be satisfied by $0 \leqq v \leqq 1$. Hence we conclude that $v^* = 0$ or $v^* = 1$. For $v = 0$, $\mu_4(1) = 0$ and (5.104) along with (5.101) cannot be satisfied for $\mu_3(1) \leqq 0$ and $0 \leqq u \leqq 1$. Hence

$$v^* = 1. \tag{5.105}$$

Substituting (5.105) into (5.101) we have

$$u = \tfrac{1}{10}\left[1 + 4\mu_1(1) - 4\mu_2(1)\right] \tag{5.106}$$

and $u^* = \frac{1}{10}$, 0, or 1.

Suppose $u=1$. Then $\mu_1(1)=0$ and (5.106) becomes

$$1 = \frac{1-4\mu_2(1)}{10},$$

which cannot be satisfied with $\mu_2(1) \geqq 0$. Conversely, suppose $u=0$. Then $\mu_2(1)=0$ and (5.106) becomes

$$0 = \frac{1+4\mu_1(1)}{10},$$

which cannot be satisfied for $\mu_1(1) \geqq 0$. Thus the min-max point for Player 1 is

$$(u^*, v^*)_1 = \left(\tfrac{1}{10}, 1\right). \tag{5.107}$$

To determine the min-max solution for Player 2 define

$$\begin{aligned} L_2 = & -3x_2 + v - \lambda_1(2)(-3x_1 + u - u^2 - x_1 v) \\ & -\lambda_2(2)(-2x_2 + v - v^2 - 2x_2 u) - \mu_1(2)u - \mu_2(2)(1-u) \\ & -\mu_3(2)v - \mu_4(2)(1-v), \end{aligned}$$

where $\mu_1(2), \mu_2(2) \leqq 0$ and $\mu_3(2), \mu_4(2) \geqq 0$. Then

$$\frac{\partial L_2}{\partial x_1} = 0 = \lambda_1(2)(3+v) \tag{5.108}$$

$$\frac{\partial L_2}{\partial x_2} = 0 = -3 + 2\lambda_2(2)(1+u) \tag{5.109}$$

$$\frac{\partial L_2}{\partial u} = 0 = -\lambda_1(2)(1-2u) + 2\lambda_2(2)x_2 - \mu_1(2) + \mu_2(2) \tag{5.110}$$

$$\frac{\partial L_2}{\partial v} = 0 = 1 + \lambda_1(2)x_1 - \lambda_2(2)(1-2v) - \mu_3(2) + \mu_4(2). \tag{5.111}$$

For $u \geqq 0$ and $v \geqq 0$, (5.108) and (5.109) yield

$$\lambda_1(2)=0$$

$$\lambda_2(2)=\frac{3}{2(1+u)}.$$

Using these results, we find that (5.110) and (5.111) become

$$0=\frac{3x_2}{1+u}-\mu_1(2)+\mu_2(2) \tag{5.112}$$

$$0=1-\frac{3(1-2v)}{2(1+u)}-\mu_3(2)+\mu_4(2). \tag{5.113}$$

Solving for x_2 from (5.94) yields

$$x_2=\frac{v-v^2}{2(1+u)} \tag{5.114}$$

and (5.112) becomes

$$0=\frac{3v(1-v)}{2(1+u)^2}-\mu_1(2)+\mu_2(2). \tag{5.115}$$

From (5.113)

$$v=\tfrac{1}{2}-\tfrac{1}{3}[1-\mu_3(2)+\mu_4(2)](1+u). \tag{5.116}$$

Substituting (5.116) into (5.115), we have

$$0=\frac{3\left\{\frac{1}{4}-\frac{1}{9}[1-\mu_3(2)+\mu_4(2)]^2(1+u)^2\right\}}{2(1+u)^2}-\mu_1(2)+\mu_2(2). \tag{5.117}$$

Suppose $0<u^*<1$. Then $\mu_1(2)=\mu_2(2)=0$ and (5.115) yields $v=0$ or $v=1$. If $v=1$, then $\mu_3(2)=0$ and (5.113) yields

$$0 \leqq \mu_4(2)=-1-\frac{3}{2(1+u)},$$

which cannot be satisfied with $u\in[0,1]$. Hence $0<u^*<1$ implies $v=0$.

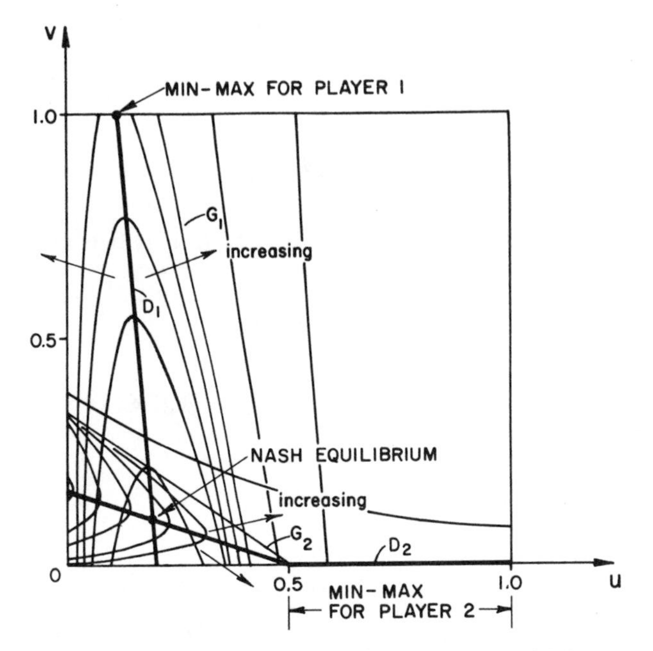

Figure 5.9. Geometry for Examples 5.4 and 5.8.

Then $\mu_3(2)=0$ and (5.113) yields

$$0 \leqq \mu_3(2)=1-\frac{3}{2(1+u)},$$

which requires $u \geqq \frac{1}{2}$. Thus $v=0$ and $u \in [\frac{1}{2}, 1)$ are (candidate) min-max points for Player 2.

Suppose $u^*=0$. Then $\mu_2(2)=0$ and (5.115) yields

$$0 \geqq \mu_1(2)=\tfrac{3}{2}v(1-v),$$

which can only be satisfied, with $v \in [0,1]$, at $v=0$ or at $v=1$. If $v=0$, then $\mu_4(2)=0$ and (5.113) cannot be satisfied at $(u,v)=(0,0)$ with $\mu_3(2) \geqq 0$. Similarly, at $(u,v)=(0,1)$, $\mu_1(2)=0$ and (5.113) cannot be satisfied with $\mu_4(2) \geqq 0$. Hence $u^* \neq 0$.

Suppose $u^*=1$. Then $\mu_1(2)=0$ and (5.115) yields

$$0 \geqq \mu_2(2)=-\tfrac{3}{8}v(1-v),$$

which is satisfied for all $v \in [0,1]$. Suppose $v \in (0,1]$. Then $\mu_3(2)=0$ and (5.113), evaluated at $u=1$, yields

$$0 \leqq \mu_4(2) = -1 + \tfrac{3}{4}(1-2v),$$

which cannot be satisfied with $v \in (0,1]$. Hence $u=1$ implies $v=0$ and therefore $\mu_4(2)=0$. The point $(u,v)=(1,0)$ is a candidate min-max point for player 2, with $\mu_3(2)=\frac{1}{4}$ from (5.113).

Thus the min-max points for Player 2 are $\{(u,v) \in E^2 \mid v=0,\ u \in [\frac{1}{2},1]\}$, as illustrated in Figure 5.9.

5.6 MIN-MAX COUNTERPOINT SOLUTIONS

The Nash equilibrium solution concept is suited for the case in which each player knows the cost functions and constraints for all of the other players and in which no cooperation is possible. On the other hand, the min-max solutions for Player i can be determined by Player i based only on knowledge of his own cost criterion and of the constraints on the other players. In particular, knowledge of the other players' cost criteria is not required. For this reason it is reasonable to expect that in many games each player may select one of his min-max controls.

In this section we introduce a min-max counterpoint solution concept. Under this solution concept all players except Player i are assumed to play one of their min-max controls. Player i has complete knowledge of the cost functions and constraints for the other players (i.e., he can determine their min-max controls) and seeks to minimize his own cost, assuming that the other players select min-max controls.

DEFINITION 5.5 A point $u=(u^1,\ldots,u^r) \in \Omega$ is a *composite min-max control* if and only if for each $j=1,\ldots,r$, u^j is a min-max control for Player j.

Let u^i denote the control for Player i and let v denote the composite control for the remaining players. Let $x=\xi(u^i,v)$ denote the solution to $g(x,u^i,v)=0$.

DEFINITION 5.6 A point $\tilde{u}=(\tilde{u}^i,\tilde{v}) \in \Omega$ is a *min-max counterpoint for Player i* if and only if $\tilde{v}$ is a composite min-max control for the remaining players and

$$G_i(\xi(\tilde{u}),\tilde{u}) \leqq G_i(\xi(u^i,\tilde{v}),u^i,\tilde{v})$$

for all $u^i \in E^{s_i}$ such that $(u^i,\tilde{v}) \in \Omega$.

Note that composite min-max controls $\tilde{v}$ and min-max counterpoints $\tilde{u}$ may not be unique, if they exist at all.

Let D_i denote the rational reaction set for Player i. Then we have the following lemma.

Lemma 5.5 A point $\tilde{u}=(\tilde{u}^i, \tilde{v})\in\Omega$ is a min-max counterpoint for Player i if and only if $\tilde{v}$ is a composite min-max control for the other players and

$$(\tilde{u}^i, \tilde{v})\in D_i.$$

Proof The lemma follows immediately from Definition 5.6. ■

Example 5.9 Consider the problem discussed in Examples 5.4 and 5.8, illustrated in Figure 5.9, and defined by

$$G_1(\cdot)=-5x_1+u$$

$$G_2(\cdot)=-3x_2+v$$

$$0=g_1(\cdot)=-3x_1+u-u^2-x_1v$$

$$0=g_2(\cdot)=-2x_2+v-v^2-2x_2u$$

$$0\leqq u\leqq 1$$

$$0\leqq v\leqq 1.$$

The min-max points for Players 1 and 2, respectively, are

$$(u^*, v^*)_1=\left(\tfrac{1}{10}, 1\right)$$

$$(u^*, v^*)_2\in\left\{(u,0)\mid u\in\left[\tfrac{1}{2}, 1\right]\right\}.$$

Thus the composite min-max control is

$$(u, v)=\left(\tfrac{1}{10}, 0\right),$$

which is neither a min-max point for either player nor a Nash equilibrium. Indeed, from Figure 5.9, the composite min-max point is not on either player's rational reaction set.

Solving for x_1 and x_2 from $g_1(\cdot)=0$ and $g_2(\cdot)=0$ and using direct substitution, we have

$$G_1(u,v)=\frac{5u^2-2u+uv}{3+v}$$

$$G_2(u,v)=\frac{3v^2-v+2uv}{2(1+u)}.$$

Assuming Player 2 uses his min-max control $v=0$, the min-max counterpoint control for Player 1 is $u=\frac{2}{5}$, which minimizes $G_1(u,0)$. Thus the min-max counterpoint for Player 1 is

$$(\tilde{u},\tilde{v})_1=\left(\tfrac{2}{5},0\right),$$

which lies on the rational reaction set for Player 1.

Assuming Player 1 plays his min-max control $u=\frac{1}{10}$, the min-max counterpoint control for Player 2 is $v=\frac{2}{15}$, which minimizes $G_2(\frac{1}{10},v)$. Thus the min-max counterpoint for Player 2 is

$$(\tilde{u},\tilde{v})_2=\left(\tfrac{1}{10},\tfrac{2}{15}\right),$$

which lies on the rational reaction set for Player 2.

For completely regular local min-max counterpoints we have the following theorem.

Theorem 5.3 If $\tilde{u}=(\tilde{u}^i,\tilde{v})\in\Omega$ is a completely regular local min-max counterpoint for Player i for the game (5.1)–(5.3) and if $\tilde{x}=\xi(\tilde{u})$ is the solution to $g(x,\tilde{u})=0$, then there exist vectors $\lambda(i)\in E^n$ and $\mu(i)\in E^q$ such that

$$\frac{\partial L_i[\tilde{x},\tilde{u},\lambda(i),\mu(i)]}{\partial x}=0, \tag{5.118}$$

$$\frac{\partial L_i[\tilde{x},\tilde{u},\lambda(i),\mu(i)]}{\partial u^i}=0 \tag{5.119}$$

$$g(\tilde{x},\tilde{u})=0 \tag{5.120}$$

$$h(\tilde{x},\tilde{u})\geqq 0 \tag{5.121}$$

$$\mu^T(i)h(\tilde{x},\tilde{u})=0 \tag{5.122}$$

$$\mu(i)\geqq 0 \tag{5.123}$$

where

$$L_i[x, u, \lambda(i), \mu(i)] \triangleq G_i(x, u) - \lambda^T(i)g(x, u) - \mu^T(i)h(x, u) \tag{5.124}$$

and $\tilde{v}$ is a composite min-max control for the other players.

Proof From Definition 5.6 $\tilde{v}$ is a composite min-max control for all of the players except i. Given $\tilde{v}$, let $x = \tilde{\xi}(u^i) \triangleq \xi(u^i, \tilde{v})$ denote the solution to $g(x, u^i, \tilde{v}) = 0$ and define $\tilde{G}_i(x, u^i) \triangleq G_i(x, u^i, \tilde{v})$. Then the theorem follows from Theorem 4.1. ■

A relationship between min-max counterpoints, Nash equilibrium points, and composite min-max points is given by the following theorem.

Theorem 5.4 There exists a Nash equilibrium point $\hat{u}$, a composite min-max point u for all of the players, and min-max counterpoints $\tilde{u}(i)$ for each player $i = 1, \ldots, r$ such that

$$\hat{u} = u = \tilde{u}(i) \qquad \forall i = 1, \ldots, r$$

if and only if there exist min-max counterpoints $\tilde{u}(i)$ for each player $i = 1, \ldots, r$ such that

$$\tilde{u}(i) = \tilde{u}(j) \qquad \forall i, j = 1, \ldots, r.$$

Proof Suppose there exist min-max counterpoints $\tilde{u}(i)$, $i = 1, \ldots, r$ that are identical, so that $\tilde{u} \triangleq \tilde{u}(i) = \tilde{u}(j)$ for all $i, j = 1, \ldots, r$. Then from Lemma 5.5 $\tilde{u}(i) \in D_i$ for each $i = 1, \ldots, r$, where D_i is the rational reaction set for Player i. Hence $\hat{u} \triangleq \tilde{u} \in \cap_{i=1}^{r} D_i$ is a Nash equilibrium point according to Lemma 5.1. To show that $u \triangleq \tilde{u}$ is also a composite min-max point, we note that, if $\tilde{u}(i) = [\tilde{u}^1(i), \ldots, \tilde{u}^r(i)]$ is a min-max counterpoint for Player i, then, for each $k = 1, \ldots, r$, $k \neq i$, $\tilde{u}^k(i)$ is a min-max control for Player k. Hence $u = \tilde{u}(i) = \tilde{u}(j)$ for all $i, j = 1, \ldots, r$ implies that $\tilde{u} = (\tilde{u}^1, \ldots, \tilde{u}^r)$ is a composite min-max point since $\tilde{u}^i$ is a min-max control Player i for each $i = 1, \ldots, r$.

Conversely, suppose there exists a Nash equilibrium point $\hat{u} = (\hat{u}^1, \ldots, \hat{u}^r)$ such that $\hat{u} = u$, where u is a composite min-max point. For each

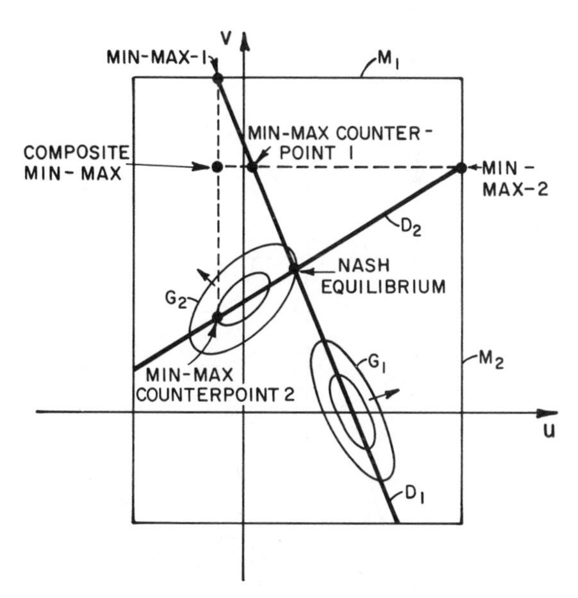

Figure 5.10. The geometry of Nash equilibrium point, composite min-max points, and min-max counterpoints.

$i=1,\ldots,r$ consider the point $\tilde{u}(i)=[\hat{u}^i,\tilde{v}(i)]=\hat{u}=u$, where $\tilde{v}(i)=(u^1,\ldots,u^{i-1},u^{i+1},\ldots,u^r)$ is the composite min-max control for the remaining players. Since $\hat{u}=u$ and $\hat{u}^i$ is a Nash equilibrium control, the conditions of Definition 5.6 are satisfied. Hence for each $i=1,\ldots,r$, $\tilde{u}(i)\triangleq\hat{u}=u$ is a min-max counterpoint for Player i. ■

Figure 5.10 illustrates the geometry of a two-player game for the case in which the Nash equilibrium point, the composite min-max point, and the min-max counterpoints do not coincide. Figure 5.11 illustrates a two-player game in which the min-max counterpoints are identical to each other and to the Nash equilibrium point and the composite min-max point.

Note that, even if the conditions of Theorem 5.4 hold, the min-max points for the respective players need not coincide. However, we do have the following result.

Theorem 5.5 If there exist min-max points $u^*(i)$ for each player $i=1,\ldots,r$ such that

$$u^*(1)=\cdots=u^*(r)\triangleq u^*, \tag{5.125}$$

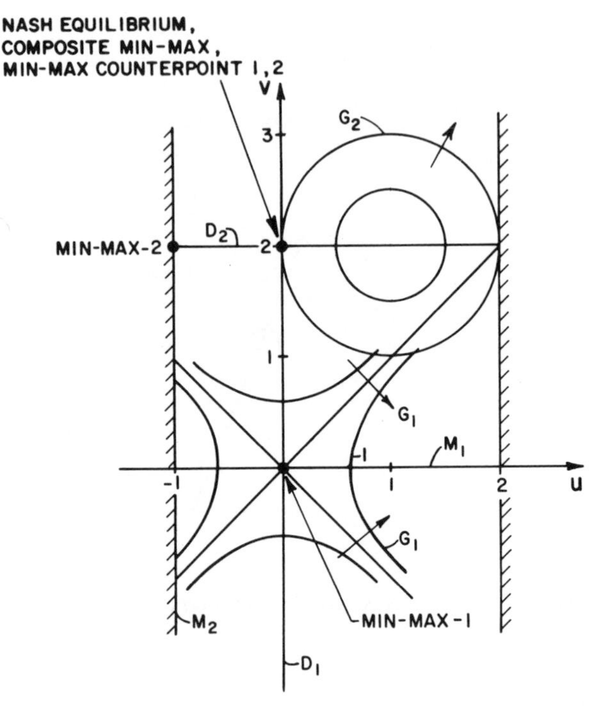

Figure 5.11. Coincident Nash equilibrium point, composite min-max points, and min-max counterpoints.

then the common point u^* is a Nash equilibrium point, a composite min-max point, and a min-max counterpoint for each player $i=1,\ldots,r$.

Proof From (5.125) and Definition 5.5 u^* is a composite min-max point. Moreover, from (5.125) and Definition 5.6, $\tilde{u}(i) \triangleq u^*$ is a min-max counterpoint for Player i for each $i=1,\ldots,r$. Hence from Theorem 5.4 u^* is also a Nash equilibrium point. ■

5.7 PARETO-MINIMAL SOLUTIONS

The Nash equilibrium solution addresses the situation in which all players act independently, pursuing their own interests. In particular, cooperation among the players is not considered. In contrast to the Nash equilibrium solution, the min-max solution considers the case where all players except one cooperate against the lone remaining player.

In this section we consider the case where cooperation among all of the players is possible. It is assumed that each player helps the others up to the point of disadvantage to himself. This is the Pareto-minimal (cooperative) solution concept and the results of Chapter 3 apply.

Let $x=\xi(u)$ denote the solution to (5.2) given $u\in\Omega$, where Ω is defined by (5.3).

DEFINITION 5.7 A point $u^\circ\in\Omega$ is a *Pareto-minimal solution* if and only if there does *not* exist a $u\in\Omega$ such that

$$G_i(\xi(u),u)\leqq G_i(\xi(u^\circ),u^\circ)$$

for all $i\in\{1,\dots,r\}$ and

$$G_j(\xi(u),u)<G_j(\xi(u^\circ),u^\circ)$$

for some $j\in\{1,\dots,r\}$. More compactly, $u^\circ\in\Omega$ is a Pareto-minimal control if and only if there does *not* exist a $u\in\Omega$ such that

$$G(\xi(u),u)\leqslant G(\xi(u^\circ),u^\circ)$$

where $G(\cdot)=[G_1(\cdot),\dots,G_r(\cdot)]^T$.

Proof The conditions follow directly from Theorem 3.3. ■

Theorem 5.6 If $u^\circ\in\Omega$ is a regular local Pareto-minimal solution for the game (5.1)–(5.3) and if $x^\circ=\xi(u^\circ)$ is the corresponding solution to $g(x,u^\circ)=0$, then there exist vectors $\eta\in E^r$, $\lambda\in E^n$, $\mu\in E^q$, where $n\geqslant 0$ and $\mu\geqq 0$, such that

$$\frac{\partial L(x^\circ,u^\circ,\eta,\lambda,\mu)}{\partial x}=0 \tag{5.126}$$

$$\frac{\partial L(x^\circ,u^\circ,\eta,\lambda,\mu)}{\partial u}=0 \tag{5.127}$$

$$g(x^\circ,u^\circ)=0 \tag{5.128}$$

$$\mu^T h(x^\circ,u^\circ)=0 \tag{5.129}$$

$$h(x^\circ,u^\circ)\geqq 0 \tag{5.130}$$

where

$$L(x,u,\eta,\lambda,\mu)\triangleq\eta^T G(x,u)-\lambda^T g(x,u)-\mu^T h(x,u). \tag{5.131}$$

Since (5.126), (5.127), (5.129), and (5.131) are homogeneous in (η, λ, μ), the condition $\eta=(\eta_1,\ldots,\eta_r)\geqslant 0$ may be replaced by

$$\eta_i \geqq 0 \qquad i=1,\ldots,r \tag{5.132}$$

$$\sum_{i=1}^{r} \eta_i = 1. \tag{5.133}$$

Example 5.10 Consider a two-player game with

$$G_1(\cdot)=(u-2)^2+(v-1)^2$$

$$G_2(\cdot)=(u-1)^2+\tfrac{1}{2}(v-2)^2$$

where Player 1 controls $u\in E^1$ and Player 2 controls $v\in E^1$ with $0\leqq u\leqq 4$ and $0\leqq v\leqq 4$. We examine the Pareto-minimal solutions as well as the Nash equilibrium solution and the min-max solutions for each player. For the Pareto-minimal solutions define

$$L=\eta_1\left[(u-2)^2+(v-1)^2\right]+\eta_2\left[(u-1)^2+\tfrac{1}{2}(v-2)^2\right]$$
$$-\mu_1(4-u)-\mu_2 u-\mu_3(4-v)-\mu_4 v.$$

Then from Theorem 5.6

$$\frac{\partial L}{\partial u}=0=2\eta_1(u-2)+2\eta_2(u-1)+\mu_1-\mu_2 \tag{5.134}$$

$$\frac{\partial L}{\partial v}=0=2\eta_1(v-1)+\eta_2(v-2)+\mu_3-\mu_4. \tag{5.135}$$

Using (5.132) and (5.133) and solving for u and v from (5.134) and (5.135), respectively, we have

$$u=1+\eta_1+\tfrac{1}{2}(\mu_2-\mu_1)$$

$$v=\frac{2-\mu_3+\mu_4}{1+\eta_1}$$

where $0\leqq\eta_1\leqq 1$. These conditions, along with (5.129) and (5.130), and $\mu\geqq 0$, require $0<u<4$ and $0<v<4$. Thus $\mu=0$ and the Pareto-minimal control set is

$$P=\left\{(u,v)\in E^2 \mid v=\frac{2}{u},\ 1\leqq u\leqq 2\right\}.$$

By inspection the Nash equilibrium point is

$$(\hat{u},\hat{v})=(2,2),$$

which is the intersection of the rational reaction sets

$$D_1=\{(u,v)\mid u=2,\ 0\leqq v\leqq 4\}$$

$$D_2=\{(u,v)\mid v=2,\ 0\leqq u\leqq 4\}.$$

The min-max points for Players 1 and 2 may also be determined by inspection, yielding

$$(u^*,v^*)_1=(2,4)$$

$$(u^*,v^*)_2=(4,2).$$

Lines of constant cost and the various solution points are shown in Figure 5.12. Note that the Nash equilibrium point and the two min-max points are not Pareto-minimal. However, the Nash equilibrium and the composite min-max point are coincident and, from Theorem 5.4, both are identical to the min-max counterpoints.

The necessary conditions for Nash equilibria, min-max points, and Pareto-minimal points have been stated in terms of various scalar-valued L functions. In the analysis of any particular game, however, it is often convenient to work in terms of just the Pareto-minimal L function, which can be specialized to any specific case by appropriate adjustment of the multipliers.

The Pareto-minimal L function is

$$L=\eta_1 G_1(x,u)+\cdots+\eta_r G_r(x,u)-\lambda^T g(x,u)-\mu^T h(x,u)$$

and the necessary conditions for a Pareto-minimum are given by (5.126)–(5.131) in Theorem 5.6. The Nash equilibrium conditions of Theorem 5.1 and the min-max conditions of Theorem 5.2 can be obtained from the conditions of Theorem 5.6 by setting the appropriate components of η to 0 or 1 and by evaluating partial derivatives of $L(\cdot)$ using appropriate choices

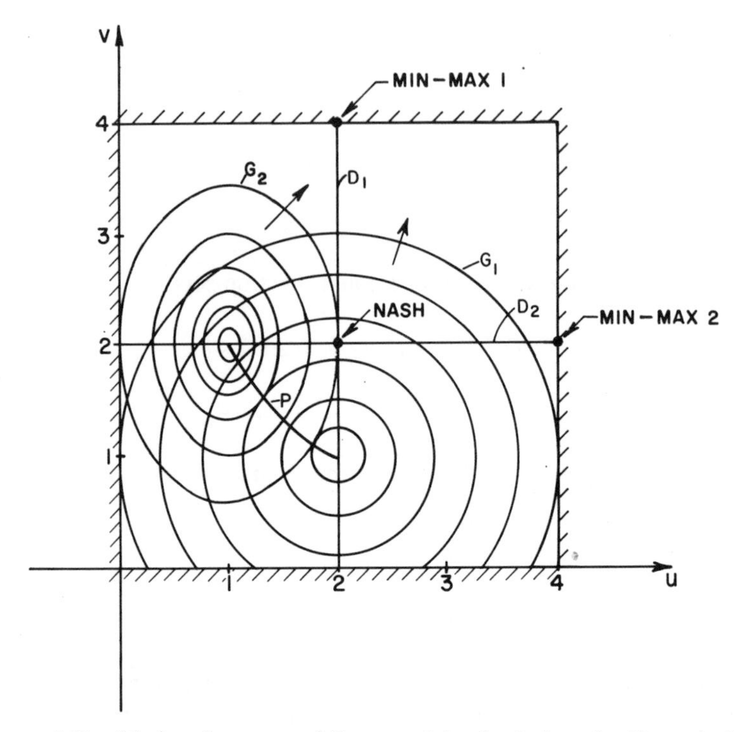

Figure 5.12. Nash, min-max, and Pareto-minimal solutions for Example 5.10.

for λ and μ. For example, the necessary conditions for a min-max control $u^*=(u^{i*}, v^*)$ for Player i can be obtained from (5.126)–(5.131) with

$$\eta_j = \delta_{ij}$$

where δ_{ij} is the Kronecker delta:

$$\delta_{ij} = \begin{cases} 1 & \text{if } i=j \\ 0 & \text{if } i \neq j. \end{cases}$$

Conditions (5.126)–(5.131) are evaluated using $\lambda(1)$ and $\mu(1)$ with $\mu(1) \geqq 0$ and using $\partial/\partial u^i$ in (5.127). The same conditions are evaluated again using $\lambda(2)$ and $\mu(2)$ with $\mu(2) \leqq 0$ and using $\partial/\partial v$ in (5.127). The Nash equilibrium conditions

$$\frac{\partial L_i}{\partial u^i} = 0 \qquad i=1,\ldots,r,$$

where $L_i(\cdot)$ is defined by (5.33) with $u=(u^1,\ldots,u^r)$, can be obtained from

$$\frac{\partial L}{\partial u^i}=0 \qquad i=1,\ldots,r$$

by evaluating L at $\lambda=\lambda(i)$ and $\mu=\mu(i)$ with

$$\eta_j=\delta_{ij}$$
$$\mu(i)\geqq 0$$

for $i=1,\ldots,r$.

5.8 BARGAINING SETS

Consider an r-player game, defined by (5.1)–(5.3), in which cooperation among the players is possible. The Pareto-minimal solution concept is well suited for this situation. In general, the Pareto-minimal controls form a set of possible controls rather than a single point. The negotiation problem is concerned with selecting composite controls in the Pareto-minimal set that are agreeable to all players. It is assumed that no player will agree to a solution whose cost to him is worse than some specified *bargaining limit cost*, such as a min-max cost.

Let $u=(u^1,\ldots,u^r)$ denote the composite control vector and let $G(\cdot)=[G_1(\cdot),\ldots,G_r(\cdot)]^T$ denote the composite cost vector. Let $x=\xi(u)$ denote the solution to (5.2) generated by $u\in\Omega$ and let $P\subseteq\Omega$ denote the Pareto-minimal control set.

DEFINITION 5.8 A point $u\in\Omega$ is a *bargaining point for Player i* if and only if:

(i) $u\in P$

and

(ii) $G_i(\xi(u),u)\leqq G_i^*$,

where G_i^* is a specified bargaining limit cost for Player i. The set of all bargaining points for Player i is denoted by $P_i^*\subseteq\Omega$. The (composite) *bargaining set* for the game is

$$P^*=\{u\in P\,|\,G(\xi(u),u)\leqq G^*\}$$

where $G^*=(G_1^*,\ldots,G_r^*)$ and G_i^* is the bargaining limit cost for Player i, $i=1,\ldots,r$.

Example 5.11 Consider a two-player game with

$$G_1(\cdot)=u^2-v \tag{5.136}$$

$$G_2(\cdot)=(u-2)^2+(v-1)^2 \tag{5.137}$$

where Player 1 controls $u\in E^1$ and Player 2 controls $v\in E^2$, with $0\leqq u\leqq 3$ and $0\leqq v\leqq 3$, and the bargaining limit cost for each player is his min-max cost. Define

$$L=\eta_1[u^2-v]+\eta_2[(u-2)^2+(v-1)^2]$$
$$-\mu_1 u-\mu_2(3-u)-\mu_3 v-\mu_4(3-v).$$

Then at a Pareto-minimal point

$$\frac{\partial L}{\partial u}=0=2\eta_1 u+2\eta_2(u-2)-\mu_1+\mu_2 \tag{5.138}$$

$$\frac{\partial L}{\partial v}=0=-\eta_1+2\eta_2 v-2\eta_2-\mu_3+\mu_4. \tag{5.139}$$

Rather than examine all $\eta\geqslant 0$, we may take $0\leqq\eta_1\leqq 1$ and $0\leqq\eta_2\leqq 1$ with

$$\eta_1+\eta_2=1. \tag{5.140}$$

Then (5.138) and (5.139) may be written as

$$0=2u-4\eta_2-\mu_1+\mu_2 \tag{5.141}$$

$$0=2\eta_2 v-\eta_2-1-\mu_3+\mu_4 \tag{5.142}$$

where $\eta_2\in[0,1]$. Since the control constraints are state-independent and decoupled, the min-max point for Player 1 may be determined from (5.140)–(5.142) by setting $\eta_1=1$ and $\eta_2=0$, yielding

$$u=\tfrac{1}{2}(\mu_1-\mu_2) \tag{5.143}$$

$$1=\mu_4-\mu_3 \tag{5.144}$$

where $\mu_1\geqq 0$, $\mu_2\geqq 0$, $\mu_3\leqq 0$, and $\mu_4\leqq 0$.

If $0<v<3$, then (5.144) cannot be satisfied with $\mu_3=\mu_4=0$. Thus $v=0$

or $v=3$. For $v=3$, $\mu_3=0$ and, again, (5.144) cannot be satisfied with $\mu_4 \leqq 0$. Thus, at the min-max point for Player 1, $v=0$ and, for $\mu_4=0$, (5.144) is satisfied with $\mu_3=-1$. Similarly, if $0<u<3$, then (5.143) cannot be satisfied with $\mu_1=\mu_2=0$. Hence $u=0$ or $u=3$. For $u=3$, $\mu_1=0$ and (5.143) cannot be satisfied with $\mu_2 \geqq 0$.

Thus the min-max point for Player 1 is

$$(u^*, v^*)_1 = (0,0) \tag{5.145}$$

with cost

$$G_1^* = 0. \tag{5.146}$$

In a similar fashion, the min-max point for Player 2 may be determined from (5.140) and (5.142) by setting $\eta_1=0$ and $\eta_2=1$, yielding

$$u = 2 + \frac{\mu_1 - \mu_2}{2} \tag{5.147}$$

$$v = 1 + \frac{\mu_3 - \mu_4}{2} \tag{5.148}$$

where, now, $\mu_1 \leqq 0$, $\mu_2 \leqq 0$, $\mu_3 \geqq 0$, and $\mu_4 \geqq 0$.

For $0<u<3$, $\mu_1=\mu_2=0$ and (5.147) yields $u=2$. But this provides a local minimum of $G_2(\cdot)$ for any given v, instead of a local maximum. Hence $u=0$ or $u=3$, both of which are candidates. For $u=0$, $\mu_2=0$ and (5.147) yields $\mu_1=-4<0$. For $u=3$, $\mu_1=0$ and (5.147) yields $\mu_2=-2$. For $0<v<3$, $\mu_3=\mu_4=0$ and (5.148) yields $v=1$. For $v=0$, $\mu_4=0$ and (5.148) cannot be satisfied with $\mu_3 \geqq 0$. Similarly, for $v=3$, $\mu_3=0$ and (5.148) cannot be satisfied with $\mu_4 \geqq 0$. Thus the candidate min-max points for Player 2 are $(u, v)=(0,1)$ and $(u, v)=(3,1)$. Comparing $G_2(\cdot)$ at these two points, we find that the min-max point for Player 2 is

$$(u^*, v^*)_2 = (0,1) \tag{5.149}$$

with cost

$$G_2^* = 4. \tag{5.150}$$

The Pareto-minimal points are determined from (5.141) and (5.142) with $\mu \geqq 0$. For interior points, with $\mu=0$, we have from (5.141)

$$u = 2\eta_2 \tag{5.151}$$

where $\eta_2 \in [0,1]$. Combining (5.142) and (5.151) yields

$$v = \tfrac{1}{2} + \frac{1}{u}. \tag{5.152}$$

Thus if there were no constraints on u and v, the Pareto-minimal points would be $u \in [0,2]$ and v given by (5.152).

The presence of constraints on u and v induces Pareto-minimal points in addition to those given by (5.152) with $u \in [0,2]$. Suppose $v=0$. Then $\mu_4=0$ and (5.142) cannot be satisfied with $\eta_2 \in [0,1]$ and $\mu_3 \geqq 0$. Hence $v=3$, $\mu_3=0$, and (5.142) yields

$$\mu_4 = 1 - 5\eta_2. \tag{5.153}$$

For $\mu_4 \geqq 0$, (5.153) and $\eta_2 \in [0,1]$ imply

$$0 \leqq \eta_2 \leqq \tfrac{1}{5}. \tag{5.154}$$

Suppose $0 < u \leqq 3$. Then $\mu_1 = \mu_2 = 0$, and for $v=3$, (5.141) and (5.154) imply $u \in (0, \tfrac{2}{5}]$. The point $u=0$, $v=3$ is also a candidate, since for $\mu_2=0$ (5.141) can be satisfied by $\mu_1 = \eta_2 = 0$.

Thus the Pareto-minimal set P is given by

$$P = \left\{ (u,v) \in E^2 \mid v = \bar{v}(u),\ 0 \leqq u \leqq 2 \right\}$$

where

$$\bar{v}(u) = \begin{cases} \tfrac{1}{2} + \dfrac{1}{u} & u \in \left[\tfrac{2}{5}, 2\right] \\ 3 & u \in \left[0, \tfrac{2}{5}\right]. \end{cases} \tag{5.155}$$

In Figure 5.13 $P = \overline{AE}$.

The bargaining set for Player 1, based on min-max cost, is given by all $(u,v) \in P$ such that

$$G_1(\cdot) = u^2 - v \leqq G_1^* = 0. \tag{5.156}$$

For $v=3$ and $u \in [0, \tfrac{2}{5}]$, (5.156) is satisfied. For $u \in [\tfrac{2}{5}, 2]$, (5.156) becomes

$$u^2 - \frac{1}{u} - \tfrac{1}{2} \leqq 0,$$

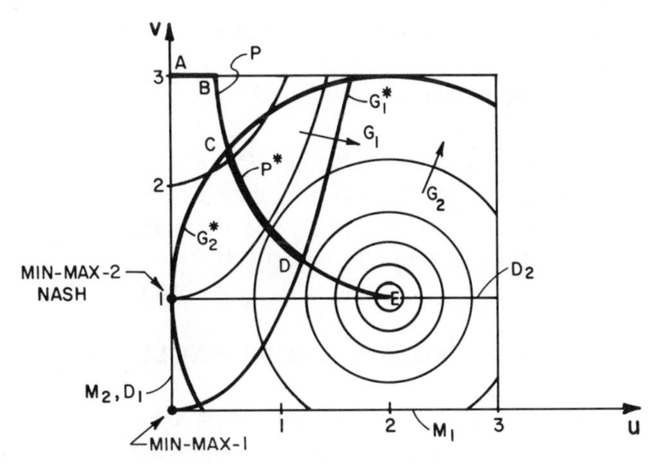

Figure 5.13. Geometry for Example 5.11.

which implies

$$u \leqq 1.1654.$$

Thus the bargaining set for Player 1 is

$$P_1^* = \{(u, v) \in P \,|\, 0 \leqq u \leqq 1.1654\}.$$

The bargaining set for Player 2, based on min-max cost, is given by all $(u, v) \in P$ such that

$$G_2(\cdot) = (u-2)^2 + (v-1)^2 \leqq G_2^* = 4. \tag{5.157}$$

For $v=3$, (5.157) cannot be satisfied for $u \in [0, \frac{2}{5}]$. For $u \in [\frac{2}{5}, 2]$, using (5.155), we have

$$(u-2)^2 + \frac{(2-u)^2}{2u^2} \leqq 4,$$

which implies

$$u \geqq 0.5367.$$

Thus the bargaining set for Player 2 is

$$P_2^* = \{(u, v) \in P \,|\, 0.5367 \leqq u \leqq 2\}$$

and the composite bargaining set, illustrated in Figure 5.13, is

$$P^* = \{(u,v) \in P \mid 0.5367 \leqq u \leqq 1.1654\}.$$

In Figure 5.13, $P_1^* = \overline{AD}$, $P_2^* = \overline{CE}$, and $P^* = \overline{CD}$. For reference the Nash equilibrium point $(\hat{u}, \hat{v}) = (0, 1)$ is also shown in Figure 5.13 along with the rational reactions sets D_1 and D_2 and the maximizing reactions sets M_2 and M_1 for Players 1 and 2, respectively.

5.9 STACKELBERG LEADER-FOLLOWER SOLUTIONS

In this section we consider generalizations of the Stackelberg solution concept (Von Stackelberg, 1952) as applied to nonzero-sum games (Simaan and Cruz, 1973). Under the Stackelberg concept one player (the Leader) announces his control $u \in E^m$ first. Then the remaining players (the Followers) announce their composite control $v \in E^s$ simultaneously.

Consider a game with $r+1$ players. The Leader, Player 0, controls $u \in E^m$ and seeks to

$$\text{minimize } G_0(x, u, v). \tag{5.158}$$

Given u, the Followers, Players $1, \ldots, r$, control $v^i \in E^{s_i}$, respectively, and seek to

$$\text{“minimize” } G(x, u, v) \tag{5.159}$$

under some optimality concept (to be specified), where $G(\cdot) = [G_1(\cdot), \ldots, G_r(\cdot)]^T$ is an r-vector and $v = (v^1, \ldots, v^r) \in E^s$. The controls u and v and the state $x \in E^n$ are subject to the inequality constraints

$$h(x, u, v) \geqq 0. \tag{5.160}$$

The state $x = \xi(u, v)$ is determined implicitly by the equality constraints

$$g(x, u, v) = 0 \tag{5.161}$$

where we assume

$$\left| \frac{\partial g(x, u, v)}{\partial x} \right| \neq 0 \tag{5.162}$$

in a ball about a solution point (x, u, v), yet to be defined.

The function $G_0(\cdot): E^n \times E^m \times E^s \to E^1$ is C^1. The functions $G(\cdot): E^n \times E^m \times E^s \to E^r$, $g(\cdot): E^n \times E^m \times E^s \to E^n$, and $h(\cdot): E^n \times E^m \times E^s \to E^q$ are C^2 in x and v and C^1 in u.

It is assumed that the Leader knows the optimality concept of the Followers. The mode of play considered here is the following. The Leader will announce $u \in E^m$. Then, based on their optimality concept, denoted by "min," the Followers will select $v(u) \in E^s$ to

$$\text{"min" } G[\xi(u,v),u,v] \tag{5.163}$$

subject to (5.160) and (5.161). The Leader, knowing the optimality concept that generates the Followers' response $v(u)$, seeks u to

$$\min G_0[\xi[u,v(u)],u,v(u)], \tag{5.164}$$

also subject to (5.160) and (5.161). The notation $v(u)$ represents a mapping relating u to the Followers' response. The Leader may be able to use knowledge of the response of the Followers to his advantage in minimizing his own cost. The Followers may also benefit from having a Leader in that they do not have to guess what the Leader will do. As a consequence neither the Leader nor the Followers necessarily have an advantage. That is, having the first (or last) move may or may not be advantageous.

Example 5.12 Consider a two-player zero-sum game in which Player 1 controls $u \in E^1$ and seeks to minimize

$$G_1(\cdot) = (u-v)^2$$

while Player 2 controls $v \in E^1$ and seeks to minimize

$$G_2(\cdot) = -(u-v)^2$$

with $-1 \leqq v \leqq 3$.

Since this example is a two-player zero-sum game, the Nash equilibrium solution and the min-max solutions (if they exist) are identical. However, the min-max and Nash equilibrium solutions do not exist for this example, since the rational reaction sets do not intersect (see Example 5.7).

Suppose Player 2 were the Leader. Then, for any $v \in [-1,3]$ that Player 2 announces, Player 1, as the Follower, would rationally choose $u=v$, producing a cost of $G_2=0$ for Player 2 as Leader and $G_1=-G_2=0$ for Player 1 as the Follower.

Conversely, suppose Player 1 were the Leader. If Player 1 were to announce $u \geqq 1$, Player 2 as the Follower could rationally select $v=-1$, producing a cost of $G_1=(u+1)^2>0$ for the Leader and $G_2=-(u+1)^2<0$ for the Follower. Similarly, if Player 1 as the Leader were to announce $u \leqq 1$, Player 2 as the Follower could rationally select $v=3$, producing a cost of $G_1=(u-3)^2>0$ for the Leader and $G_2=-(u-3)^2<0$ for the Follower.

Thus, in this example, we see that each player would prefer to have the other player as the Leader.

Consider the Leader's problem of predicting the Followers' response to $u \in E^m$. In the simplest case the response is given by a C^1 function $v(u)$ and the function $v(\cdot)$ is itself known to the Leader. Then (5.164) is a scalar minimization problem in u subject to the constraints $h[\xi[u, v(u)], u, v(u)] \geqq 0$ and $g[\xi[u, v(u)], u, v(u)]=0$. In the more general case that we consider here, the Leader does not have *a priori* knowledge in the form of a C^1 response function $v(\cdot)$. Indeed, the response mapping $v(\cdot)$ may not be single-valued, or continuous, much less C^1.

Various optimality concepts are possible for the Followers. In the absence of cooperation among the Followers they might, for example, play Nash equilibrium among themselves, min-max against Follower i, or composite min-max. If cooperation among the Followers is possible, they might first agree to restrict themselves to responses $v \in P(u)$, where $P(u)$ denotes the Pareto-minimal control set for $G(\cdot)$ given u. Since $P(u)$ is generally a set of points, the Followers might further agree to a negotiated response that minimizes or maximizes some additional criterion (such as the Leader's cost) subject to the constraint $v \in P(u)$.

Example 5.13 The Leader controls $u \in E^1$ and the Followers control $v \in E^2$. The Leader seeks to minimize the scalar-valued criterion

$$G_0(u, v)=u^2+v_1-v_2$$

and the Followers seek a Pareto-minimal point for the vector-valued criteria

$$G(\cdot)=[G_1(\cdot), G_2(\cdot)]^T$$

$$=[(v_1-2)^2+v_2^2, v_1^2+(v_2-2)^2]^T.$$

Without additional information, the response set for the Followers is their Pareto-minimal control set

$$P=\{v \in E^2 \mid v_2=2-v_1, 0 \leqq v_1 \leqq 2\}.$$

Thus a degree of freedom exists for the Followers in selecting a Pareto-minimal response. The Leader's problem of predicting the Followers' response is not well defined until we specify an additional optimality concept and a cost criterion for the Followers in selecting a particular Pareto-minimal point $v(u)$ as their response to $u \in E^1$.

For this example suppose the Followers seek to maximize the Leader's cost, subject to the constraint that the Followers must be Pareto-minimal among themselves. In other words, the Followers seek $v \in E^2$ to

$$\max_v \left\{ G_0(u,v) = u^2 + v_1 - v_2 \right\}$$

subject to

$$v_1 + v_2 - 2 = 0$$
$$0 \leqq v_1 \leqq 2.$$

Then the Leader seeks $u \in E^1$ to

$$\min_u \max_{0 \leqq v_1 \leqq 2} \left\{ \tilde{G}_0(u,v) = u^2 + 2v_1 - 2 \right\}.$$

The solution is $(u, v_1, v_2) = (0, 2, 0)$ and the resulting costs are $(G_0, G_1, G_2) = (2, 0, 8)$. Note that, if the Followers drop the restriction that their response be Pareto-minimal among themselves, then they could force the Leader's cost to be as large (or as small) as they desire.

Necessary conditions can be developed for a variety of Stackelberg Leader-Follower optimality concepts. For the remainder of this chapter we consider the case of Nash equilibrium Followers. This includes the special case, considered by Simaan and Cruz (1973), of a single Follower with a single scalar-valued criteria. In addition, we consider problems that may be subject to equality and inequality constraints.

Let $x = \xi(u, v)$ denote the solution to (5.161) and let

$$\Omega = \left\{ (u,v) \in E^m \times E^s \mid h[\xi(u,v), u, v] \geqq 0 \right\}. \tag{5.165}$$

DEFINITION 5.9 The Follower's *Nash response set* $\hat{D}$ is the set of all $(u, v) = (u, v^1, \ldots, v^r) \in \Omega \subseteq E^m \times E^s$ such that for each $i = 1, \ldots, r$,

$$G_i(\xi(u,v), u, v) \leqq G_i[\xi[u, \bar{v}(i)], u, \bar{v}(i)] \tag{5.166}$$

for all $\bar{v}(i) \in E^s$ such that $[u, \bar{v}(i)] \in \Omega$, where $\bar{v}(i) = (\bar{v}^1, \ldots, \bar{v}^r)$ is identical to v except for the controls of Follower i. Put differently, $\bar{v}^j = v^j$, $j = 1, \ldots, r$, $j \neq i$.

DEFINITION 5.10 A point $(\hat{u}, \hat{v}) \in \Omega$ is a *Stackelberg solution for the Leader with Nash Followers* if and only if $(\hat{u}, \hat{v}) \in \hat{D}$ and

$$G_0(\xi(\hat{u}, \hat{v}), \hat{u}, \hat{v}) \leq G_0(\xi(u, v), u, v) \tag{5.167}$$

for all $(u, v) \in \hat{D}$. For a local solution (5.167) must hold for all $(u, v) \in B \cap \hat{D}$ for some ball $B \subset E^m \times E^s$ centered at $(\hat{u}, \hat{v})$.

Example 5.14 The Leader controls $u \in E^1$ and seeks to minimize

$$G_0(\cdot) = (u-2)^2 + v^2.$$

A single Follower controls $v \in E^1$ and seeks to minimize the scalar-valued criterion

$$G_1(\cdot) = (u-v)^2.$$

The Follower's response to $u \in E^1$ is $v(u) = u$ and his response set is

$$\hat{D} = \{(u, v) \in E^2 | v = u\},$$

as illustrated in Figure 5.14. Thus the Leader selects $\hat{u} = 1$ to minimize

$$G_0[u, v(u)] = (u-2)^2 + u^2$$

and the Stackelberg solution for the Leader with a minimizing Follower is $(\hat{u}, \hat{v}) = (1, 1)$. If the two players change roles then the Stackelberg solution is $(\hat{u}, \hat{v}) = (2, 2)$, which is also the Nash solution.

If $v = (v^1, \ldots, v^r)$ is a completely regular local Nash equilibrium solution for the Followers given $u \in E^m$ and x is the corresponding state satisfying (5.160)–(5.162), then from Theorem 5.1 there exist vectors $\lambda(i) \in E^n$, $\mu(i) \in E^q$, $i = 1, \ldots, r$ such that

$$h(x, u, v) \geq 0 \tag{5.168}$$

$$g(x, u, v) = 0 \tag{5.169}$$

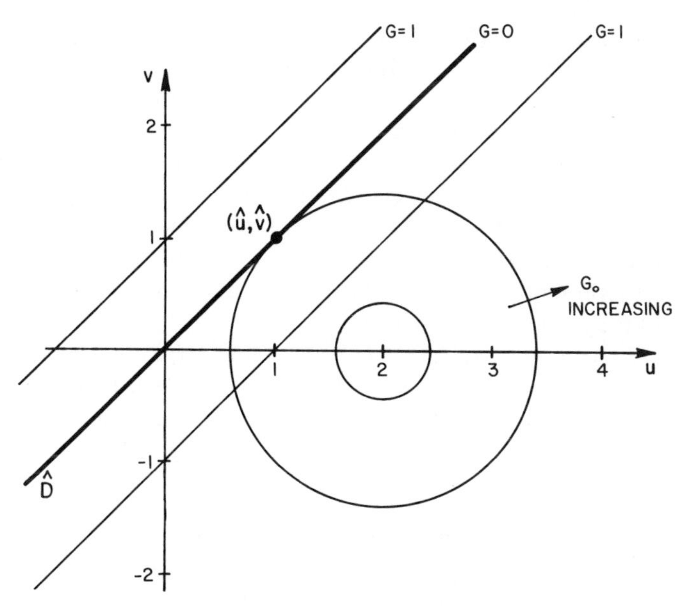

Figure 5.14. The Stackelberg point for Example 5.14.

and for each $i=1,\ldots,r$

$$\frac{\partial \hat{L}_i(x,u,v,\lambda(i),\mu(i)]}{\partial x}=0 \tag{5.170}$$

$$\frac{\partial \hat{L}_i(x,u,v,\lambda(i),\mu(i)]}{\partial v^i}=0 \tag{5.171}$$

$$\mu_j(i)\geqq 0 \qquad \text{for all } j\in\hat{Q}_i \tag{5.172}$$

where $\hat{Q}=\{j\,|\,h_j(x,u,v)=0\}$, $\mu_j(i)=0$ for all $j\notin\hat{Q}$, and

$$\hat{L}_i[x,u,v,\lambda(i),\mu(i)] \triangleq G_i(x,u,v)-\lambda^T(i)g(x,u,v)-\sum_{j\in\hat{Q}}\mu_j(i)h_j(x,u,v). \tag{5.173}$$

In view of Definition 5.10 one approach to developing necessary conditions for the Leader would be to consider the problem of minimizing $G_0(\xi(u,v),u,v)$ subject to the constraints $(u,v)\in\hat{D}\subset E^m\times E^s$. Lack of *a priori* knowledge of $\xi(\cdot)$ would not pose any difficulty since $\xi(\cdot)$ exists implicitly as a result of (5.162). But $\hat{D}$ does create a problem. In the first place $\hat{D}$ is not known *a priori*. Instead $\hat{D}$ is expressed, through the Nash equilibrium necessary conditions for the Followers, in terms of x, u, v, and r pairs of Lagrange multiplier vectors $\lambda(i)$, $\mu(i)$, $i=1,\ldots,r$; that is, $\hat{D}\subset E^m \times E^s$ is expressed indirectly in terms of variables in a space Y of dimension $n+m+s+r(n+q)$. More importantly, however, even if $\hat{D}$ could be expressed directly as a set in $E^m \times E^s$, $\hat{D}$ generally may be disconnected or possess "corners." In other words, $\hat{D}\subset E^m \times E^s$ is generally not regular. But the indirect representation of $\hat{D}$ in the higher dimensional space Y is regular and allows the application of Theorem 2.2.

Let $y=[x, u, v, \lambda(1), \mu(1),\ldots, \lambda(r), \mu(r)]$ and let Y be the set of all y satisfying (5.168)–(5.173). Let $\hat{Y}$ be the set of all y such that v is a regular local Nash equilibrium solution for the Followers for $u\in E^m$ with corresponding state x and vectors $\lambda(i)$ and $\mu(i)$ satisfying (5.168)–(5.173). Then $\hat{Y}\subseteq Y$.

DEFINITION 5.11 A point $(\hat{u}, \hat{v})\in\Omega$, with corresponding $\hat{y}=[\xi(\hat{u}, \hat{v}), \hat{u}, \hat{v}, \lambda(1), \mu(1),\ldots, \lambda(r), \mu(r)]\in Y$ satisfying (5.168)–(5.173), is a *completely regular* local Stackelberg solution for the Leader with Nash Followers if and only if:

(i) $\hat{y}\in\hat{Y}$;
(ii) there exists a ball B centered at $\hat{y}$ such that $B\cap\hat{Y}=B\cap Y$;
(iii) $\hat{y}$ minimizes $G_0(\cdot)$ on $B\cap Y$;
(iv) $(\hat{u}, \hat{v})$ is a completely regular point of Ω.

If a point $(\hat{u}, \hat{v})$ is a completely regular local Stackelberg solution for the Leader with Nash Followers, then locally every solution to (5.168)–(5.173) corresponds to a completely regular local Nash equilibrium solution for the Followers. Note that, from the continuity of $h(\cdot)$, any inactive constraints corresponding to (5.168) can be kept inactive by keeping the ball B in Definition 5.11 sufficiently small.

As a consequence of Lemma 2.10 we have the following lemma.

Lemma 5.6 If $(\hat{u}, \hat{v})\in\Omega$ is a completely regular local Stackelberg solution for the Leader with Nash Followers, with corresponding $\hat{y}\in Y$, then $\hat{y}$ is a completely regular point of $B\cap\hat{Y}=B\cap Y$, where B is the ball in Definition 5.10 and Y is defined by the solutions to (5.168)–(5.173).

Proof In view of (5.162) $x=\xi(u, v)$ is uniquely determined locally from (5.169) on $B\cap Y$ and $\xi(\cdot)$ is C^1. Hence (5.169) is a regular constraint at $(\hat{u}, \hat{v})$. The constraints (5.170), (5.171), and (5.172) are linear in $\lambda(i)$ and $\mu_j(i)$, $j\in\hat{Q}$, on $B\cap Y$ and are therefore regular constraints at $(\hat{x}, \hat{u}, \hat{v})$

from Lemma 2.10. The remaining constraints (5.168) and (5.169) define Ω. Thus, since $(\hat{u}, \hat{v})$ is a completely regular point of Ω, the corresponding point $\hat{y}$ is a completely regular point of $B \cap Y$. ■

Theorem 5.7 If $(\hat{u}, \hat{v}) = (\hat{u}, \hat{v}^1, \ldots, \hat{v}^r) \in \Omega$ is a completely regular local Stackelberg solution for the Leader with Nash Followers, and $\hat{x} = \xi(\hat{u}, \hat{v})$ is the solution to $g(x, \hat{u}, \hat{v}) = 0$, then for each Follower, $i = 1, \ldots, r$, there exist vectors $\lambda(i) \in E^n$ and $\mu(i) \in E^q$ such that

$$\frac{\partial L_i}{\partial x} = 0 \tag{5.174}$$

$$\frac{\partial L_i}{\partial v^i} = 0 \tag{5.175}$$

$$g(\hat{x}, \hat{u}, \hat{v}) = 0 \tag{5.176}$$

$$h(\hat{x}, \hat{u}, \hat{v}) \geqq 0 \tag{5.177}$$

$$\mu^T(i) h(\hat{x}, \hat{u}, \hat{v}) = 0 \tag{5.178}$$

$$\mu(i) \geqq 0 \tag{5.179}$$

where

$$L_i[x, u, v, \lambda(i), \mu(i)] \triangleq G_i(x, u, v) - \lambda^T(i) g(x, u, v) - \mu^T(i) h(x, u, v) \tag{5.180}$$

with (5.174) and (5.175) evaluated at $(x, u, v) = (\hat{x}, \hat{u}, \hat{v})$. Furthermore, for the Leader, there exist vectors $\lambda(0) \in E^n$, $\mu(0) \in E^q$, $\alpha(i) \in E^n$, $\beta(i) \in E^{s_i}$, and $\rho(i) \in E^q$, $i = 1, \ldots, r$, such that

$$\frac{\partial L_0}{\partial x} = 0 \tag{5.181}$$

$$\frac{\partial L_0}{\partial u} = 0 \tag{5.182}$$

$$\frac{\partial L_0}{\partial v} = 0 \tag{5.183}$$

$$\mu^T(0) h(x, u, v) = 0 \tag{5.184}$$

$$\mu(0) \geqq 0 \tag{5.185}$$

and for each $i=1,\ldots,r$

$$\frac{\partial L_0}{\partial \lambda(i)}=0 \tag{5.186}$$

$$\frac{\partial L_0}{\partial \mu_j(i)}=0 \qquad j\in\hat{Q} \tag{5.187}$$

$$\rho^T(i)\mu(i)=0 \tag{5.188}$$

$$\rho(i)\geqq 0 \tag{5.189}$$

$$\rho^T(i)h(\hat{x},\hat{u},\hat{v})=0, \tag{5.190}$$

where

$$\hat{Q}=\{j\,|\,h_j(\hat{x},\hat{u},\hat{v})=0\} \tag{5.191}$$

and

$$L_0[x,u,v,\lambda(1),\mu(1),\alpha(1),\beta(1),\rho(1),\ldots,\lambda(r),\mu(r),\alpha(r),\beta(r),\rho(r)]$$

$$\triangleq G_0(x,u,v)-\lambda^T(0)g(x,u,v)-\mu^T(0)h(x,u,v)$$

$$-\sum_{i=1}^{r}\left\{\frac{\partial L_i}{\partial x}\alpha(i)+\frac{\partial L_i}{\partial v^i}\beta(i)+\rho^T(i)\mu(i)\right\} \tag{5.192}$$

where (5.181)–(5.183), (5.186), and (5.187) are evaluated at $(x,u,v)=(\hat{x},\hat{u},\hat{v})$.

Proof Conditions (5.174)–(5.180) follow directly from Theorem 5.1 for Nash Followers. Conditions (5.181)–(5.192) follow directly from Theorem 2.2 applied to the problem of minimizing $G_0(\cdot)$ subject to (5.168)–(5.173). ■

Note that (5.188)–(5.190) imply

$$\rho_j(i)=0 \qquad \text{for all } j\notin\hat{Q} \tag{5.193}$$

and

$$\rho_j(i)=0 \qquad \text{for all } j\in\hat{Q} \text{ with } \mu_j(i)>0. \tag{5.194}$$

Thus the only possible nonzero $\rho_j(i)$ corresponds to $h_j(\hat{x}, \hat{u}, \hat{v})=0$ with $\mu_j(i)=0$, in which case $\rho_j(i)\geqq 0$.

Example 5.15 The Leader controls $u\in E^1$ and seeks to minimize

$$G_0(u,v)=\tfrac{1}{2}(u-1)^2+\tfrac{1}{2}(v-2)^2.$$

A single Follower controls $v\in E^1$ and seeks to minimize

$$G(u,v)=\tfrac{1}{2}(u-v)^2,$$

where we drop the subscript "1" since we have only a single Follower. Both players are subject to

$$h(u,v)=2-u^2-v^2\geqq 0.$$

For the Follower

$$L(\cdot)=\tfrac{1}{2}(u-v)^2-\mu(2-u^2-v^2),$$

and the Follower's necessary conditions (5.175) and (5.177)–(5.180) are

$$\frac{\partial L}{\partial v}=v-u+2\mu v=0 \tag{5.195}$$

$$h(\cdot)=2-u^2-v^2\geqq 0 \tag{5.196}$$

$$\mu\geqq 0 \tag{5.197}$$

$$\mu(2-u^2-v^2)=0. \tag{5.198}$$

If $h(\cdot)>0$, then $\mu=0$, (5.195) yields

$$v=u, \tag{5.199}$$

and $h(\cdot)>0$ requires

$$|u|<1. \tag{5.200}$$

If $h(\cdot)=0$, then (5.195) yields

$$v=\frac{u}{1+2\mu} \tag{5.201}$$

where $\mu \geqq 0$, and $h(\cdot)=0$ yields

$$2-u^2-v^2=2-u^2-\frac{u^2}{(1+2\mu)^2}=0.$$

Thus for $h(\cdot)=0$

$$u^2=\frac{2}{1+1/(1+2\mu)^2} \tag{5.202}$$

and $\mu \geqq 0$ implies

$$1 \leqq |u| < \sqrt{2}\,.$$

The Follower's reaction set, as illustrated Figure 5.15, is

$$\hat{D}=\left\{(u,v)\in E^2 \mid v=\hat{v}(u),\, |u| \leqq \sqrt{2}\right\}$$

where

$$\hat{v}(u)=\begin{cases} u & \text{if} \quad |u| \leqq 1 \\ \sqrt{2-u^2} & \text{if} \quad 1 \leqq u \leqq \sqrt{2} \\ -\sqrt{2-u^2} & \text{if} \quad -\sqrt{2} \leqq u \leqq -1 \end{cases} \tag{5.203}$$

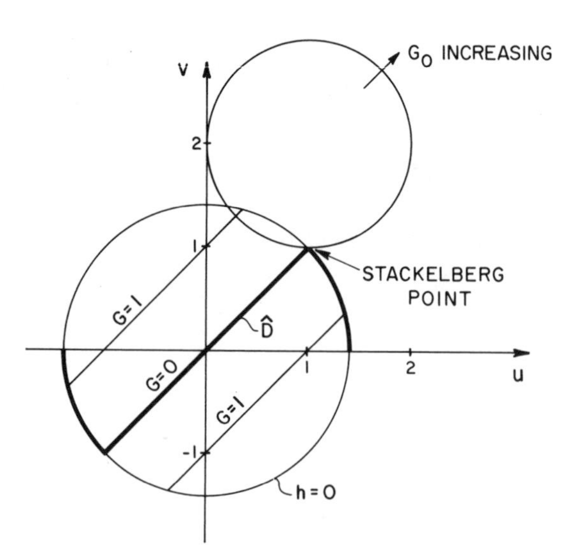

Figure 5.15. The Stackelberg point for Example 5.15.

and we have used (5.201) and $\mu \geqq 0$ to determine the sign of v on $h(\cdot)=0$.

It is interesting to note that the Follower's optimal reaction $v=0$ for $u=\pm\sqrt{2}$ does not satisfy the necessary conditions (5.195) on $h(\cdot)=0$ unless $\mu=+\infty$ is allowed. However, these results do occur in the limit as $\mu \to +\infty$ in (5.201) and (5.202). The peculiarity stems from the fact that at $u=\pm\sqrt{2}$, $v=0$ is not a (one-dimensional) regular point of the Follower's control constraint set

$$V(u)=\{v \in E^1 \mid h(u,v) \geqq 0\}.$$

To verify this, we note that only $v=0$ satisfies $h(\cdot) \geqq 0$ at $u^2=2$. Thus the tangent cone $T_v \subseteq E^1$ to $V(u)$ at $u=\pm\sqrt{2}$ is $T_v=\{0\}$. But

$$\frac{\partial h}{\partial v}=-2v=0$$

at $v=0$ and the set $\hat{T}_v$ of all $e^v \in E^1$ satisfying

$$\frac{\partial h}{\partial v} e^v \geqq 0$$

at $v=0$ is $\hat{T}_v=E^1 \neq T_v$. Note, however, that the points $(u,v)=(\pm\sqrt{2},0)$ are regular points of $\Omega \subset E^2$. Thus (u,v) is regular, but not completely regular.

Turning now to the Leader, we define

$$\begin{aligned} L_0 &= G_0-\mu_0 h-\frac{\partial L}{\partial v}\beta-\rho\mu \\ &= \tfrac{1}{2}(u-1)^2+\tfrac{1}{2}(v-2)^2-\beta[v(1+2\mu)-u] \\ &\quad -\mu_0(2-u^2-v^2)-\rho\mu. \end{aligned}$$

Then the additional necessary conditions are

$$\frac{\partial L_0}{\partial u}=u-1+\beta+2u\mu_0=0 \tag{5.204}$$

$$\frac{\partial L_0}{\partial v}=v-2-\beta(1+2\mu)+2v\mu_0=0 \tag{5.205}$$

$$\mu_0 \geqq 0 \tag{5.206}$$

$$\mu_0(2-u^2-v^2)=0. \tag{5.207}$$

Furthermore, if $h(\cdot)=2-u^2-v^2=0$, we also impose conditions (5.187)–(5.190), that is

$$\frac{\partial L_0}{\partial \mu}=-2v\beta-\rho=0 \tag{5.208}$$

$$\rho \geqq 0 \tag{5.209}$$

$$\rho\mu=0. \tag{5.210}$$

Suppose $h(\cdot)>0$. Then $\mu=\mu_0=0$, $u=v$ with $|u|<1$, and (5.204) and (5.205) yield $\beta=1-u=u-2$, which implies $u=\frac{3}{2}$ and violates $|u|<1$. Therefore we have $h(\cdot)=0$ and

$$u^2=\frac{2}{1+1/(1+2\mu)^2} \tag{5.211}$$

$$v=\frac{u}{1+2\mu}. \tag{5.212}$$

where $\mu \geqq 0$. In addition (5.208)–(5.210) apply.

Suppose $0<\mu<+\infty$. Then (5.211) and (5.212) imply $v\neq 0$. Therefore (5.208)–(5.210) yield $\rho=\beta=0$ and (5.204) and (5.205) become

$$u(1+2\mu_0)-1=0 \tag{5.213}$$

$$v(1+2\mu_0)-2=0. \tag{5.214}$$

Conditions (5.212)–(5.214) yield

$$2=\frac{1}{1+2\mu},$$

which cannot be satisfied with $0<\mu<+\infty$.

Thus $\mu=0$ or $\mu=+\infty$. For $\mu=0$ (5.195) yields $u=v$ and $h(\cdot)=0$ implies

$$u=v=\pm 1. \tag{5.215}$$

Subtracting (5.205) from (5.204) produces

$$(u-v)(1+2\mu_0)+1+2\beta(1+\mu)=0.$$

Thus for $\mu=0$ and $u=v$ we have

$$\beta=-\tfrac{1}{2}$$

and (5.208)–(5.210) yield

$$v=\rho\geqq 0. \tag{5.216}$$

Thus from (5.215) and (5.216) we have the candidate Stackelberg point

$$u=v=1. \tag{5.217}$$

with $\beta=-\frac{1}{2}$, $\rho=1$, $\mu=0$ and, from (5.204), $\mu_0=\frac{1}{4}$. The points $v=0$, $u=\pm\sqrt{2}$, corresponding to $\mu=+\infty$, are also candidates. However, from

$$G_0(\pm\sqrt{2},0)=2+\tfrac{1}{2}(\pm\sqrt{2}-1)^2$$

and

$$G_0(1,1)=\tfrac{1}{2},$$

we conclude that the Stackelberg solution for the Leader is $\hat{u}=\hat{v}=1$.

5.10 EXERCISES

For Exercises 5.1 through 5.5 determine the rational reaction set for each player and all candidates for local Pareto-minimal, Nash, and min-max solutions. Reduce the Pareto-minimal solution set to the bargaining set, using each player's largest min-max cost as his bargaining limit cost. All games are two-player games and all players have scalar controls. Player 1 has cost G_1 with control u_1, and so on.

5.1 $G=[u_1+u_2, 0.2u_1-0.5u_2]^T$, $h=[u_1, 1-u_1, u_2, 1-u_2]^T$.

5.2 $G=[u_1^2+u_2^2, (u_1-0.5)^2+(u_2-0.5)^2]^T$, $h=[u_1, 1-u_1, u_2, 1-u_2]^T$.

5.3 $G=[u_1^2+u_2^2, -u_1^2-u_2]^T$, $h=[u_1+1, 1-u_1, u_2+1, 1-u_2]^T$.

5.4 $G=[-x_1+x_2, -x_1+x_3]^T$, $g=[x_1-u_1-u_2, x_2-u_1^2, x_3-u_2^2]^T$, $h=[u_1, 1-u_1, u_2, 2-u_2]^T$.

5.5 Rework Exercise 5.1 with the inequality conditions replaced by the single inequality $(\frac{1}{2})^2-(u_1-\frac{1}{2})^2-(u_2-\frac{1}{2})^2\geqq 0$.

5.6 Two people are jointly going to build a box without top or bottom. Let x=length, u_1=height, and u_2=width. The amount of lumber available is fixed at 60 ft^2. The first person can choose the height and wants to maximize the volume enclosed by the box. The second person can choose the width and wants to maximize the moment of inertia per unit mass of air enclosed by the box about the x-axis through the center of the mass. Assume all lumber is used and $\frac{1}{2} \leqq u_1 \leqq 10, \frac{1}{2} \leqq u_2 \leqq 10$. Determine the Nash, min-max, and Pareto-minimal solutions.

5.7 (The Utility Game) The Customer sees electrical use as a tool and convenience and requests that lines be attached to his home. He has a fixed cost k_1 associated with no electrical use. Let u_1=consumption. The actual consumption is a partly negative cost for him (he does not have to burn wood, he has the added convenience, etc.). Let u_2=price of electricity per unit consumption. Naturally, the Customer must pay for this consumption, with cost=$u_1 u_2$. Careful study shows that in the town of Noscut the Customer's total cost is given by

$$\text{Customer cost} = k_1 - 2u_1 + u_1 u_2.$$

The Customer controls consumption u_1. The Utility desires to make a profit selling electricity to the Customer. It has fixed cost k_2 due to the maintenance of a plant that has a maximum production capability (production=consumption) of $u_{1\,max}=1$. It has a cost associated with producing electricity and, finally, it makes its profit by selling the electricity to the Customer. For Noscut the Utility's cost is given by

$$\text{Utility cost} = k_2 + \alpha u_1 - u_1 u_2.$$

The Utility controls price u_2, with $\alpha = 1$. It is apparent that there is no theoretical limit to u_2. Thus, while $0 \leqq u_1 \leqq 1$, we have $0 \leqq u_2 < \infty$. Determine if there exist price-consumption values that *both* the Customer and Utility will ultimately select. Suppose that the coefficient α in the Utility cost function were 2.1 instead of 1. What would you do if you were the Customer?

5.8 In Biology it is maintained that evolution *maximizes* individual "fitness" of a species. Fitness is sometimes defined to be the intrinsic growth rate of a species. Consider two noninteracting species 1 and 2 with fitness described by

$$\text{fitness of } N_1 \stackrel{\triangle}{=} \frac{\dot{N}_1}{N_1} = k_1 - N_1$$

$$\text{fitness of } N_2 \stackrel{\triangle}{=} \frac{\dot{N}_2}{N_2} = k_2 - N_2$$

where N_1 and N_2 are population levels for each species and $(\dot{\ })$ is the time derivative of ( ). It is clear that, without interaction, the equilibrium populations are $N_1 = k_1$ and $N_2 = k_2$. Consider now placing these two species together with each at its equilibrium number. Species 2 has the option of "invading" species 1's territory and Species 1 has the option of "repelling" the invasion. For this case fitness of each species is given by

$$\frac{\dot{N}_1}{N_1} = k_1 - N_1 - \underset{①}{u_2 N_2} + \underset{②}{u_1 u_2 N_2} - \underset{③}{u_1}$$

$$\frac{\dot{N}_2}{N_2} = k_2 - N_2 + \underset{④}{u_2} - \underset{⑤}{\alpha_1 u_1 u_2}$$

where u_2 = fraction of time N_2 spends in N_1's territory
u_1 = fraction of time N_1 spends repelling N_2
① = loss of fitness due to invasion
② = regain of fitness due to repelling
③ = loss of fitness due to time spent repelling
④ = gain in fitness due to invasion
⑤ = loss of gain due to repelling.

If we assume that each species is near its equilibrium population, then fitness for each is given, respectively, by

$$H_1(\cdot) = -u_2 k_2 + u_1 u_2 k_2 - u_1$$

$$H_2(\cdot) = u_2 - \alpha_1 u_1 u_2.$$

Assume $\alpha_1 = 3$ and $k_2 = 2$. Determine if species 2 will invade species 1's territory and if so, what species 1's response will be.

5.9 Two small countries are in an arms race. Country 1 currently has 10 units of arms and Country 2 has 6 units of arms. They are now negotiating for a possible disarmament agreement. Your are an arbitrator to the negotiations and represent a powerful third country. Find a reasonable solution to impose if the two countries' cost functions are given by

$$G_1=(6+u_2)-(u_1+10)+0.2u_1^2$$

$$G_2=(10+u_1)-(u_2+6)+0.2u_2^2$$

where u_1 and u_2 represent arms purchases to be made after the conference (possibly from the third country!) or arms destroyed. Each country has the economic capacity to buy four more units of arms. Thus

$$-10\leqq u_1\leqq 4$$

$$-6\leqq u_2\leqq 4.$$

5.10 The solution to Exercise 4.6 was simply the rational reaction sets for each player. Determine the Nash and Pareto-minimal solutions for this problem. The only inequality constraints are given by the fact that the equilibrium solutions for the state x_1 and x_2 must be nonnegative. Use the parameters $\beta=1.5$ ard $\gamma=1$. If you were a system manager, what limits would you set for the harvesting efforts u_1 and u_2? Consider using a compromise solution obtained by minimizing the distance between a utopia point and a bargaining set in yield space. Let the bargaining set be that portion of the Pareto-minimal set for which each players yield is greater than or equal to the Nash yields. Let the components of the utopia point be the maximum yields for each player over the bargaining set.

5.11 Rework Exercise 5.10 under *rate* harvesting, that is, where the dynamics of the system are given by

$$\dot{x}_1=x_1(1-x_1-x_2)-u_1$$

$$\dot{x}_2=\beta x_2\left(1-\gamma\frac{x_2}{x_1}\right)-u_2$$

and the yields for each fishery

$$H_1=u_1$$

$$H_2=u_2$$

are to be maximized subject to the inequality constraints that all equilibrium solutions for x_1 and x_2 obtained by setting $\dot{x}_1=\dot{x}_2=0$ must be nonnegative. Let $\beta=1.5$ and $\gamma=1$.

5.12 Play the following game with four others. There are no ground rules other than that at a certain time a sixth uninvolved party collects each player's decision by secret ballot. $G=[u_1-4u_1u_2+1,\ u_2^2-u_3u_4+1,\ -2u_3+u_3u_4+2,\ u_4-2u_4u_5+1,\ -u_5^2+u_1u_2+1]^T$, where $0\leqq u_i\leqq 1$, $i=1,\ldots,5$.

5.13 Determine the local Pareto-minimal points, the local Nash equilibrium points, and the local min-max points for each player in the following problem:

$$\text{Player 1:} \quad \min_u \{G_1 = u - (u^2 - 1)v\}$$

$$\text{Player 2:} \quad \min_v \{G_2 = v(4 - u^2)\}$$

$$|u| \leqq 3$$

$$|v| \leqq 2.$$

5.14 Determine the local Nash equilibrium points, the local min-max points for each player, and the local *internal* Pareto-minimal points for the following problem (do not use direct substitution!):

$$\text{Player 1:} \quad \min_u \{G_1 = ux\}$$

$$\text{Player 2:} \quad \min_v \{G_2 = x^2 - \tfrac{1}{2}v^2 + xv\}$$

where

$$g = x + u + v = 0$$

$$-2 \leqq u \leqq 3$$

$$-2 \leqq v \leqq 3.$$

5.15 Determine the local Nash equilibrium points, the local Pareto-minimal points, the local min-max points, and the bargaining set based on min-max bargaining limit costs for the following problem:

$$G_1 = (u-1)^2 + (v-2)^2$$

$$G_2 = (u-2)^2 + (v-1)^2$$

$$0 \leqq u \leqq \tfrac{3}{2}$$

$$\tfrac{1}{2} \leqq v \leqq 3.$$

5.16 Determine the Stackelberg solutions for Exercise 5.9 with Country 1 as the Leader and then with Country 2 as the Leader. There is no arbitrator. Contrast these results with the results of Exercise 5.9.

5.17 If $G = [f(u_1, u_2), -f(u_1, u_2)]^T$, comment on the usefulness and relationship between the various solution concepts.

Chapter Six

OPTIMALITY IN PARAMETRIC DYNAMIC SYSTEMS

6.1 INTRODUCTION

In terms of the control notation introduced in Chapter 4, a cost criterion generally depends only on the choice made for the composite control vector. The corresponding constant state vector is determined as a function of the control by means of the system equations $g(x, u)=0$. In this chapter we relax the requirement that the state vector be a constant. Instead we assume that, for a given choice of a *constant* control vector $u \in U$, the state of the system is a function of *time* t determined by means of differential equations of the form

$$\dot{x}=g(x, u) \tag{6.1}$$

where $(\dot{\ })$ denotes $d(\ )/dt$, $g(\cdot): E^n \times E^s \to E^n$ is a specified C^1 function of the state vector $x=[x_1, \ldots, x_n]^T$, and (composite) control vector $u=[u^1, \ldots, u^r]^T \in U$, where $U \subseteq E^s$ is a completely regular subset of the (composite) state-independent control space given by

$$U=\{u \in E^s \mid h(u) \geqq 0\} \tag{6.2}$$

where $h(\cdot): E^s \to E^q$ is C^1. In the absence of inequality constraints $U=E^s$.

Let x at time $t=0$ be given. Then for any finite $u \in U$ the unique solution of (6.1) is given by $x=\xi(t, u)$, where $\xi(\cdot): E^1 \times E^s \to E^n$ is C^1 (Coddington and Levinson, 1955) and satisfies

$$\frac{\partial \xi}{\partial t}=g(\xi(t, u), u) \tag{6.3}$$

for all t. Since $g(\cdot)$ is C^1 on $E^n \times E^s$, such a solution always exists. Furthermore, the solution is C^1 in the initial state $x(0)$ for a given u.

In what follows we assume that the initial state is specified. Since the state at time t, $\xi(t,u)$, depends not only on the parameter vector u but also on time, clearly the characteristics of an optimization problem are somewhat different from those of Chapter 4. However, a clear connection can be seen by considering any asymptotically stable parametric dynamical system of the form (6.1), for which the state evolves asymptotically to a constant value with time. At such an equilibrium point $g(x,u)=0$, which are the system equations of Chapter 4.

With the state of the system no longer static, the cost criteria may now be generalized so as to reflect the dynamic nature of the state. There are a number of ways in which the cost can be characterized. We consider the case where the *cumulative cost vector* at time t is defined by

$$G=\int_0^t \gamma(x,u)\,dt, \tag{6.4}$$

where $\gamma(\cdot): E^n \times E^s \to E^r$ is C^1 and is evaluated along $x=\xi(t,u)$. Then the time rate of change of the cost vector at time t may be expressed as a function of the state vector x and the control vector u as

$$\dot{G}=\gamma(x,u). \tag{6.5}$$

Given a solution to (6.1), there exists a unique solution to (6.5) given by $G=z(t,u)$, where $z(\cdot): E^1 \times E^s \to E^r$ is C^1 and satisfies

$$\frac{\partial z}{\partial t}=\gamma(\xi(t,u),u) \tag{6.6}$$

for all t, with the initial cost $z(0,u)$ set equal to the zero vector for convenience.

Note that the cost G given by (6.4) accumulates with time. There is a fundamental difference between the static optimization problems considered in previous chapters and the dynamical problems to be considered in this chapter. Since the state x evolves with time, the control that minimizes the cost at some final time generally is *not* the same as the control that minimizes $\dot{G}$ evaluated at the initial state $x(0)$ or, for that matter, at any other state attained at a latter time. For example, consider minimum time of descent to the bottom of a mountain range starting from some point slightly below a ridge line. The optimal solution may require an initial climb over the ridge line, instead of steepest descent from the initial point.

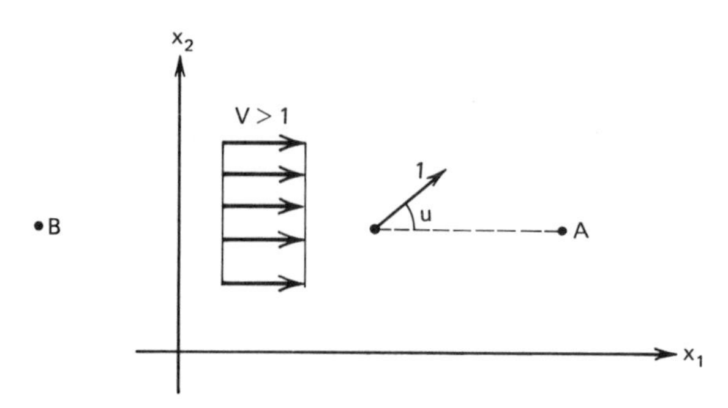

Figure 6.1. Geometry for Zermelo's problem.

Suppose that the initial and final values for the state, at some specified final time t_f, have been prescribed. We seek a constant control $u \in U$ that will not only drive the system between these points, but also minimize a cost function evaluated at the final time. Clearly, a *controllability* requirement, stating that at least one control exists that will drive the system between the initial and final points, must be met in order for an optimization problem to make sense. The problem of Zermelo (Zermelo, 1931) illustrates the controllability requirement.

Example 6.1 (Zermelo's Problem) Consider a boat moving with a velocity of unit magnitude relative to a stream of constant speed, V, and direction (along the x_1 axis of Figure 6.1). The problem is to determine a constant[†] steering angle u that will minimize the time required to go from the initial point to some final point. Clearly, if $V > 1$, then point A in Figure 6.1 can be reached by the boat (by using $u=0$ or $u=180°$) but point B cannot be reached for any value of u. Since at least one control exists for reaching point A, the minimum time problem can be solved.

Example 6.1 illustrates the fact that, under a dynamically evolving state, an optimization problem acquires two aspects: a qualitative one (controllability) and a quantitative one (optimality). The qualitative aspect is associated with the possibility of attaining one state from another. A minimum principle for determining controllability boundaries for nonlinear systems under time varying controls as developed by Grantham (1973) is discussed in Grantham and Vincent (1975). The quantitative aspect is associated with the cost obtained due to driving the system between two attainable states.

[†] This problem is usually formulated with a time varying control $u(t)$, in which case the solution must be sought using optimal control theory for dynamic systems (Leitmann, 1966).

We do not deal extensively with the qualitative aspect of the problem here. The qualitative aspects can be largely avoided by leaving the final state unspecified. However, since there are a numer of interesting problems where it is desired that the final state satisfy certain restrictions, special boundary conditions are discussed in detail in Section 6.3. We assume for all problems in this chapter that:

1 At time $t=0$ the initial state is specified.

2 For a given final time $t=t_f>0$, the final state is unrestricted.

or

2′ At an optimal final time $t=t_f>0$, the final state is constrained to be an element of a given subset of state space, which may depend on the parametric control u.

3 For all $0<t<t_f$ the state x is determined solely from (6.1).

6.2 COST PERTURBATIONS AS FUNCTIONS OF TIME

Our objective is to develop local optimality conditions for determining a parametric control $u \in U$ that optimizes either a scalar- or vector-valued cost G, where the rate of change of cost is given by (6.5), the rate of change of state is given by (6.1), and the control vector $u \in U$ is constant for the entire time duration. The control set U defined by (6.2) is a completely regular composite control set as discussed in Chapter 5 if optimality corresponds to a game concept; otherwise U corresponds to a regular control set as defined in Chapter 4. In either case, all local optimality conditions are derived from the concept of examining how the value of the cost G varies in a small neighborhood of a solution point. For dynamic systems cost perturbations in general are functions of time.

Let $u^* \in U$ be an optimal control and let

$$x=\xi(t,u^*) \tag{6.7}$$

$$G=z(t,u^*) \tag{6.8}$$

be the state vector and cost vector at time t as determined from the solutions to (6.1) and (6.5). Let B denote a ball about u^*, let $e \in T_U$ denote a vector tangent to U, where T_U is the tangent cone to U at u^*, and let $\delta u(\cdot)$ generate e, that is, $u^*+\alpha\,\delta u(\alpha) \in U$ for all sufficiently small $\alpha>0$ with $\delta u(\alpha) \to e$ as $\alpha \to 0$. Let t^* denote the terminal time (optimal or specified) and let $t^*+\alpha\,\delta t$

denote a perturbation in the terminal time. Let

$$\delta G \triangleq z[t^*+\alpha\,\delta t, u^*+\alpha\,\delta u(\alpha)]-z(t^*, u^*) \tag{6.9}$$

denote the corresponding perturbation in the cost vector, where $\delta t \in E^1$. Applying the first-order approximation theorem to (6.9) yields

$$\delta G=\frac{\partial z(t^*, u^*)}{\partial t}\alpha\,\delta t+\frac{\partial z(t^*, u^*)}{\partial u}\alpha\,\delta u(\alpha)+R(\alpha) \tag{6.10}$$

where $R(\alpha)/\alpha \to 0$ and $\delta u(\alpha) \to e$ as $\alpha \to 0$. It follows from (6.6) that

$$\delta G=\alpha\left[\gamma(\xi(t^*, u^*), u^*)\,\delta t+\frac{\partial z(t^*, u^*)}{\partial u}\delta u(\alpha)+\frac{R(\alpha)}{\alpha}\right]. \tag{6.11}$$

Equation (6.11) represents the perturbation of the cost vector as a function of time due either to a small change in the final time or a small change in control or both. Since the various solution concepts are interpreted in terms of the signs of the components of δG, we must be able to evaluate the right-hand side of (6.11). The function $\gamma(\cdot)$ is known explicitly, but the function $\partial z(\cdot)/\partial u$ is not known explicitly unless analytical solutions to (6.1) and (6.5) are available. In general, this is not the case. Nevertheless, $\partial z(t, u)/\partial u$ may be obtained indirectly from (6.3) and (6.6).

Since $g(\cdot)$ and $\gamma(\cdot)$ are C^1 and

$$\xi(t, u^*)=\xi(0, u^*)+\int_0^t g[\xi(\tau, u^*), u^*]\,d\tau$$

and

$$z(t, u^*)=\int_0^t \gamma[\xi(\tau, u^*), u^*]\,d\tau, \tag{6.12}$$

it follows from Leibniz' rule that, for a fixed initial state,

$$\frac{\partial \xi(t, u^*)}{\partial u}=\int_0^t\left[\frac{\partial g(\xi(\tau, u^*), u^*)}{\partial x}\frac{\partial \xi(\tau, u^*)}{\partial u}+\frac{\partial g(\xi(\tau, u^*), u^*)}{\partial u}\right]d\tau \tag{6.13}$$

and

$$\frac{\partial z(t, u^*)}{\partial u}=\int_0^t\left[\frac{\partial \gamma(\xi(\tau, u^*), u^*)}{\partial x}\frac{\partial \xi(\tau, u^*)}{\partial u}+\frac{\partial \gamma(\xi(\tau, u^*), u^*)}{\partial u}\right]d\tau. \tag{6.14}$$

Thus the $n \times s$ matrix

$$\Lambda(t, u^*) \triangleq \frac{\partial \xi(t, u^*)}{\partial u} \tag{6.15}$$

is the solution to the (matrix) initial value problem

$$\dot{\Lambda} = \frac{\partial g[\xi(t, u^*), u^*]}{\partial x} \Lambda + \frac{\partial g[\xi(t, u^*), u^*]}{\partial u} \tag{6.16}$$

$$\Lambda(0, u^*) = [0] \tag{6.17}$$

and the $r \times s$ matrix

$$\Gamma(t, u^*) = \frac{\partial z(t, u^*)}{\partial u} \tag{6.18}$$

is the solution to the (matrix) initial value problem

$$\dot{\Gamma} = \frac{\partial \gamma[\xi(t, u^*), u^*]}{\partial x} \Lambda + \frac{\partial \gamma[\xi(t, u^*), u^*]}{\partial u} \tag{6.19}$$

$$\Gamma(0, u^*) = [0]. \tag{6.20}$$

The initial conditions (6.17) and (6.20) correspond to a fixed initial state and zero initial cost, respectively. We call the $n \times s$ matrix $\Lambda(t, u^*)$ the *state perturbation matrix* and the $r \times s$ matrix $\Gamma(t, u^*)$ the *cost perturbation matrix*.

To employ (6.11), the state equations (6.1) and the equations (6.16) and (6.19) are integrated simultaneously from their respective initial conditions, yielding solutions $\xi(t, u^*)$, $\Lambda(t, u^*)$, and $\Gamma(t, u^*)$. Then from (6.11) and (6.18), we have (in vector form)

$$\delta G = \alpha \left[\gamma(\xi(t, u^*), u^*)\, \delta t + \Gamma(t^*, u^*)\, \delta u(\alpha) + \frac{R(\alpha)}{\alpha} \right] \tag{6.21}$$

where $\delta G = [\delta G_1, \ldots, \delta G_r]^T$ and $R = [R_1, \ldots, R_r]^T$. Note that (6.21) includes both the first-order effect of a control perturbation $u^* + \alpha\, \delta u(\alpha)$ about a nominal optimal control u^* and the first-order effect of a time perturbation $t^* + \alpha\, \delta t$ about a nominal final time t^*.

6.3 TARGET SETS AND BOUNDARY CONDITIONS

It is assumed that at time $t = 0$ the initial state is specified. Furthermore, it is assumed that the terminal boundary conditions either specify the final time with the final state unspecified or that the terminal state is restricted to some subset of the state space with the terminal time unspecified.

Consider first the case of a specified final time $t=t^*>0$, with no restrictions on the final state. Let $u^*\in U$ represent a control that satisfies an optimizing solution concept (min, min-max, Nash, Pareto, etc.), and let $u^*+\alpha\,\delta u(\alpha)\in U$ for all sufficiently small $\alpha>0$ be a perturbation in u^*, with $\delta u(\alpha)\rightarrow e\in T_U$ as $\alpha\rightarrow 0$, where T_U is the tangent cone to U at u^*. Since $\delta t=0$ at $t=t^*$ for fixed terminal time, it follows from (6.21) that

$$\delta G=\alpha\left[\Gamma(t^*,u^*)\,\delta u(\alpha)+\frac{R(\alpha)}{\alpha}\right]. \tag{6.22}$$

For sufficiently small $\alpha>0$, $\delta u(\alpha)\rightarrow e$, where e is a tangent vector to U at u^*, $R(\alpha)/\alpha\rightarrow 0$, and the signs of the components of δG are determined by the components of $\Gamma(t^*,u^*)e$, provided that these components are not zero.

In the second case the terminal time t^* is unspecified. The terminal state is not specified *a priori* but is required to lie in a regular *target* set

$$\Theta=\{x\in E^n\,|\,\theta(x,u)\leqq 0,\,u\in U\} \tag{6.23}$$

where $\theta(\cdot)\colon E^n\times E^s\rightarrow E^p$ is C^1. We seek the best terminal state $\xi(t^*,u^*)$ in Θ and this point may lie either on the boundary of Θ or on the interior of Θ, depending on the cost function specified for the problem. Targets defined in terms of systems of equality constraints, that is, $\psi(x,u)=0$, may be specified in the form (6.23) using $\theta=(\psi,-\psi)$.

For problems involving target sets (instead of a fixed terminal time with a free terminal state) the question of controllability plays a nontrivial role. For such problems we restrict the discussion to the case where some control $u^*\in U$ exists that transfers the state to the target. We further restrict the discussion to terminal points $\xi(t^*,u^*)\in\Theta$ for which small admissible perturbations in u^* $(u^*+\alpha\,\delta u(\alpha)\in U)$ and in t^* produce corresponding terminal points that also lie in Θ.

Let $u^*\in U$ be a control vector that satisfies an optimizing solution concept and let $t^*\geqq 0$ be an optimal final time at which $\xi(t^*,u^*)\in\Theta$. Let $\hat{P}=\{1\ldots\hat{p}\leqq p\}$ be the set of indices† for the active target inequality constraints at the terminal time t^* (i.e., $k\in\hat{P}$ if and only if $\theta_k(\xi(t^*,u^*),u^*)=0$) and let $\hat{\theta}(\cdot)=[\theta_1(\cdot)\cdots\theta_{\hat{p}}(\cdot)]^T$. If $\xi(t^*,u^*)\in\partial\Theta$ then $\hat{P}$ is not empty. Let $t^*+\alpha\,\delta t$ and $u^*+\alpha\,\delta u(\alpha)$ be small perturbations in t^* and u^*, respec-

†Reordered, if necessary, and possibly empty.

tively, with $u^*+\alpha\,\delta u(\alpha)\in U$ for all sufficiently small $\alpha>0$ and $\delta u(\alpha)\to e$ as $\alpha\to 0$. The perturbations δt and $\delta u(\alpha)\to e$ are related to the target inequality constraints through the requirement $\xi[t^*+\alpha\,\delta t, u^*+\alpha\,\delta u(\alpha)]\in\Theta$.

From the first-order approximation theorem applied to the active target inequalities we have

$$\hat{\theta}\left[\xi[t^*+\alpha\,\delta t, u^*+\alpha\,\delta u(\alpha)], u^*+\alpha\,\delta u(\alpha)\right]$$

$$=\hat{\theta}\left[\xi(t^*,u^*),u^*\right]+\left[\frac{\partial\hat{\theta}}{\partial x}\left[\frac{\partial\xi}{\partial t}\alpha\,\delta t+\frac{\partial\xi}{\partial u}\alpha\,\delta u(\alpha)\right]+\frac{\partial\hat{\theta}}{\partial u}\alpha\,\delta u(\alpha)\right]_{t^*,u^*}+R(\alpha)$$

where $R(\alpha)/\alpha\to 0$ as $\alpha\to 0$. Since $\hat{\theta}[\xi(t^*,u^*),u^*]=0$ and $\xi[t^*+\alpha\,\delta t, u^*+\alpha\,\delta u(\alpha)]\in\Theta$, it follows from (6.3), (6.15), and (6.23) that in the limit as $\alpha\to 0$

$$\frac{\partial\hat{\theta}\left(\xi(t^*,u^*),u^*\right)}{\partial x}\left[g(\xi(t^*,u^*),u^*)\,\delta t+\Lambda(t^*,u^*)e\right]$$
$$+\frac{\partial\hat{\theta}\left(\xi(t^*,u^*),u^*\right)}{\partial u}e\leqq 0,\qquad(6.24)$$

where $\Lambda(t^*,u^*)$ is the solution to (6.16)–(6.17).

6.4 PARAMETRIC OPTIMAL CONTROL

In this section we confine the analysis to minimizing a scalar-valued ($r=1$) cost function $G=z(t,u)$ at some (fixed or optimal) final time $t=t^*$, with G satisfying

$$\dot{G}=\gamma(x,u)\qquad(6.25)$$

where $\gamma(\cdot): E^n\times E^s\to E^1$ is C^1 and $z(\cdot)$ is the solution to (6.25) generated by a constant control $u\in U$, with $z(0,u)=0$. The state vector $x=\xi(t,u)$ satisfies (6.1) with the initial state specified. The constant control vector u is restricted to a regular state-independent subset U of E^s defined by

$$U=\{u\in E^s\,|\,h(u)\geqq 0\}\qquad(6.26)$$

where $h(\cdot): E^s\to E^q$ is C^1 in a ball about a point u^* that minimizes $G=z(t,u)$ at time $t=t^*$.

If the state equations (6.1) and the cost equation (6.25) can be integrated analytically, then the methods of Chapter 2 may be employed directly.

Example 6.2 The Lotka-Volterra model for population systems is given by

$$\frac{dN}{dt}=\frac{r}{K}N(K-N), \tag{6.27}$$

where N is the biomass of a single species population feeding on a limited resource, r is the intrinsic rate of increase, and K is the carrying capacity. This model is frequently used for the management and analysis of biological systems. For example, Goh (1980) examines its use in the management of fisheries. The model has also been used by Vincent (1977, 1978a) to examine environmental adaptation of annual plants. In this latter situation an adaptive parameter u is identified that is related to r and K in such a way that the model given in (6.27) may be written in nondimensional form as

$$\dot{x}=ux(1-ux) \tag{6.28}$$

where x is a nondimensional biomass of a plant and $(\dot{\ })$ represents a derivative with respect to a nondimensional time. Equation (6.28) represents the change in plant biomass during a given growing season. During any given growing season, $0\leqq t\leqq t^*$, a plant is restricted to a given value of the adaptive parameter $u\in U$. (For simplicity it is assumed here that $U=E^1$.) However, the value of the parameter may change from season to season and is representative of the adaptive process. It may be assumed here that the most *fit* plants are those that maximize their biomass at the end of the growing season. In this case the rate of change of cost is given by

$$\dot{G}=-ux(1-ux). \tag{6.29}$$

For a constant u (6.28) can be integrated analytically, from an initial point $x(0)$ to the fixed final time t^*, yielding

$$x(t^*)=\frac{x(0)e^{ut^*}}{1+ux(0)[e^{ut^*}-1]}. \tag{6.30}$$

Thus a necessary condition for maximizing $x(t^*)$ is obtained by setting the partial derivative of $x(t^*)$ with respect to u equal to zero at $u=u^*$ to obtain

$$e^{u^*t^*} + u^*t^* = 1 + \frac{t^*}{x(0)}, \tag{6.31}$$

where $x(0) > 0$. The value of u^* as determined from (6.31) is seen to depend upon the length of the growing season t^* and the initial biomass $x(0)$. These latter two quantities may be thought of as the environmental inputs to the system.

As illustrated in Example 6.2, the solution u^* to a parametric optimal control problem is a function of the initial state. In optimal control theory, where the control is not generally a constant, two types of control are encountered; open loop control $u^*(t)$ and closed loop (feedback) control $u^*[x]$. Our parametric optimal control $u^*[x(0)]$ does not correspond directly to either of these types of nonconstant control. It should also be noted that in optimal control theory every portion of an optimal trajectory is optimal and optimal trajectories do not intersect for autonomous systems of the form (6.1). In general, neither of these two conditions hold for the parametric optimal control system. If u^* is a parametric optimal control generating the trajectory $x^*(t)$ starting from $x^*(0)$ then u^* is not necessarily the optimal parametric control starting from other points on $x^*(t)$ unless $u^* = \text{constant}$ is also the solution to the corresponding optimal control problem.

Since the dynamical state equation could be integrated analytically in Example 6.2, a simple necessary condition was employed to determine the optimal parametric control. In many cases, an analytical solution for the state equations (6.1) is not available, and the optimal control must be obtained in a less direct manner.

For the case in which the final time t^* is specified and the final state $x(t^*)$ is unrestricted, we have the following theorem.

Theorem 6.1 Let $u^* \in U$ generate a solution $x^*(t) = \xi(t, u^*)$ to (6.1) defined on $[0, t^*]$. If u^* is a regular local minimal control for the scalar-valued cost $G = z(t^*, u)$, where t^* is a specified final time and $z(t, u^*)$ is the solution to (6.25), then there exists an $n \times s$ state perturbation matrix $\Lambda(t, u^*) \triangleq \partial\xi(t, u^*)/\partial u$, satisfying

$$\dot{\Lambda} = \frac{\partial g(x^*(t), u^*)}{\partial x}\Lambda + \frac{\partial g(x^*(t), u^*)}{\partial u}, \tag{6.32}$$

$$\Lambda(0, u^*) = [0], \tag{6.33}$$

a $1\times s$ cost perturbation matrix $\Gamma(t,u^*)\triangleq\partial z(t,u^*)/\partial u$ satisfying

$$\dot{\Gamma}=\frac{\partial\gamma(x^*(t),u^*)}{\partial x}\Lambda+\frac{\partial\gamma(x^*(t),u^*)}{\partial u} \tag{6.34}$$

$$\Gamma(0,u^*)=[0] \tag{6.35}$$

and a vector $\mu\in E^q$ such that

$$\Gamma(t^*,u^*)=\mu^T\frac{\partial h(u^*)}{\partial u} \tag{6.36}$$

$$\mu^T h(u^*)=0 \tag{6.37}$$

$$\mu\geqq 0 \tag{6.38}$$

$$h(u^*)\geqq 0. \tag{6.39}$$

Proof From (6.22) and the definition of a local minimum we have

$$\Gamma(t^*,u^*)e=\frac{\partial z(t^*,u^*)}{\partial u}e\geqq 0$$

for all $e\in T_U$, where T_U, the tangent cone to U at u^*, is the set of all $e\in E^s$ such that

$$\frac{\partial\hat{h}(u^*)}{\partial u}e\geqq 0,$$

where $\hat{h}(\cdot)=[h_1(\cdot),\ldots,h_{q^*}(\cdot)]^T$ denotes the active control inequality constraints at u^*. Thus from Farkas' lemma there exists a vector $\mu\in E^q$ satisfying (6.37) and (6.38) such that

$$\frac{\partial z(t^*,u^*)}{\partial u}=\mu^T\frac{\partial h(u^*)}{\partial u}.$$

Then (6.36) follows from this and (6.18) evaluated at (t^*,u^*). Conditions (6.32)–(6.35) follow from (6.16), (6.17), (6.19), and (6.20) evaluated at $(x^*(t),u^*)$. ■

Example 6.3 Theorem 6.1 may be used to "set up" a numerical solution for Example 6.2. From (6.28) and (6.29) we have

$$g(x,u)=-\gamma(x,u)=ux(1-ux),$$

and

$$\frac{\partial g}{\partial x}=u(1-2ux) \qquad \frac{\partial g}{\partial u}=x(1-2ux)$$
$$\frac{\partial \gamma}{\partial x}=-u(1-2ux) \qquad \frac{\partial \gamma}{\partial u}=-x(1-2ux).$$

For the scalars $\Lambda=\partial\xi/\partial u$ and $\Gamma=\partial z/\partial u$, (6.32) and (6.34) become

$$\dot{\Lambda}=(\Lambda u+x)(1-2ux) \tag{6.40}$$
$$\dot{\Gamma}=-\dot{\Lambda}.$$

Since $U=E^1$, it follows from the initial conditions $\Gamma(0)=\Lambda(0)=0$, that the necessary conditions (6.36)–(6.38) reduce to

$$\Gamma(t^*,u^*)=0.$$

Given a value for t^* and $x(0)$, we can guess a value for u^*, integrate (6.28) and (6.40) to time t^*, and then check to see if $\Gamma(t^*,u^*)=0$. If so, this yields a candidate for u^*. It is easy to show numerically that the values of u obtained by this method also satisfy (6.31).

In general, the use of Theorem 6.1 requires the numerical solution of a two-point boundary value problem. For simplicity consider the case where $U=E^s$. Then for a specified final time the necessary conditions (6.36)–(6.38) reduce to

$$\Gamma(t^*,u^*)=[0] \tag{6.41}$$

The two-point boundary value problem consists of guessing the s variables u and then iteratively integrating the $ns+n+s$ equations (6.1), (6.32), and (6.33) for various values of u, with the objective of satisfying the s terminal conditions (6.41). One conceptual approach is to select a nominal estimate u for u^* and perform the integrations. Then sequentially perturb the components of the s-vector u and repeat the integration. After the s perturbations a numerical estimate of the $s\times s$ matrix

$$\frac{\partial\Gamma(t^*,u)}{\partial u} \tag{6.42}$$

can be computed by finite differences. Then Δu could be chosen by a gradient scheme based on (6.42) so that the control $u+\Delta u$ brings the system closer to satisfying (6.41). Each iteration of such a scheme would require that a total of $(s+1)[ns+n+s]$ equations be integrated.

Now consider the same two-point boundary value constant control problem using optimal control theory. For a constant control the necessary conditions of Leitmann (1966) may be written as

$$\dot{x}=g(x,u) \qquad x(0) \text{ given} \tag{6.43}$$

$$\dot{\lambda}^T=-\lambda^T\frac{\partial g}{\partial x}-\frac{\partial \gamma}{\partial x} \tag{6.44}$$

$$\dot{\beta}^T=\frac{\partial \gamma}{\partial u}+\lambda^T\frac{\partial g}{\partial u} \qquad \beta(0)=[0] \tag{6.45}$$

with terminal condition

$$\beta(t^*)=[0], \tag{6.46}$$

where $\lambda(t)$ is an n-vector (the adjoint vector) and $\beta(t)$ is an s-vector. Note that no initial or final conditions are specified for the adjoint vector $\lambda(t)$.

Using optimal control theory, the two-point boundary value problem consists of guessing the $n+s$ variables $\lambda(0)$ and u and then interatively integrating the $2n+s$ equations (6.43)–(6.45), with the objective of satisfying the s terminal conditions (6.46). By sequentially perturbing the $n+s$ variables $\lambda(0)$ and u, a numerical estimate for the $s\times(n+s)$ matrix

$$\left[\begin{array}{c|c}\dfrac{\partial\beta(t^*)}{\partial\lambda(0)} & \dfrac{\partial\beta(t^*)}{\partial u}\end{array}\right] \tag{6.47}$$

could be computed using finite differences. Then a gradient scheme based on (6.47) could be used to produce new guesses $\lambda(0)+\Delta\lambda(0)$ and $u+\Delta u$ to bring the system closer to satisfying (6.46). Each iteration would require that a total of $[n+s+1][2n+s]$ equations be integrated.

Thus Theorem 6.1 has two advantages over the optimal control approach:

1 Theorem 6.1 has fewer initial unknowns to be guessed.

2 Theorem 6.1 requires fewer integrations for each iteration of the above two-point boundary value scheme, providing

$$s^2-s-1<2n. \tag{6.48}$$

Usually there are significantly fewer control variables than state variables. Even for $s=n$ condition (6.48) holds for $n<4$. For a more likely situation, say $s=n/3$, condition (6.48) holds for $n<22$. The computational advantages of Theorem 6.1 are even more pronounced in differential games (to be discussed), where a parametric Nash equilibrium solution in terms of optimal control theory would involve a different adjoint vector $\lambda(t)$ for each of r players, producing $rn+s$ initial conditions to be guessed, where the theorem corresponding to Theorem 6.1 for differential games only requires guesses for the s control variables.

We now develop necessary conditions similar to Theorem 6.1 for the case where the final time is unspecified and the final state is restricted to a given subset of the state space, of the form (6.23). If the state equations (6.1) and the cost equation (6.5) can be integrated analytically, then the methods of Chapter 2 may be employed directly.

Example 6.4 Determine the steering angle u that will minimize the time t required to go from the origin to the target

$$\Theta=\{(x_1,x_2)|(x_1-5)^2+(x_2-1)^2\leqq 1\} \tag{6.49}$$

for Zermelo's problem with stream speed $V=2$. The system dynamics in this case are given by

$$\dot{x}_1=2+\cos u$$

$$\dot{x}_2=\sin u.$$

Under a constant control u these differential equations are easily integrated and evaluated at time t and the result may be used to define the target in terms of t:

$$(2t+t\cos u-5)^2+(t\sin u-1)^2\leqq 1. \tag{6.50}$$

We wish to find a control u^* that will minimize t subject to (6.50). Applying Theorem 2.3 with $y=(u,t)$, the L function is given by

$$L=t+\mu\left[(2t+t\cos u-5)^2+(t\sin u-1)^2-1\right],$$

so that the necessary conditions, in addition to (6.50), are

$$0=\frac{\partial L}{\partial u}=2\mu\left[-(2t+t\cos u-5)t\sin u+(t\sin u-1)t\cos u\right]$$

$$0=\frac{\partial L}{\partial t}=1+2\mu\left[(2t+t\cos u-5)(2+\cos u)+(t\sin u-1)\sin u\right]$$

$$\mu\geqq 0$$

$$\mu\left[(2t+t\cos u-5)^2+(t\sin u-1)^2-1\right]=0.$$

From $\partial L/\partial t=0$ it follows that $\mu\neq 0$. In other words, there is no solution on the interior of the target. Thus (6.50) is satisfied as an equality at the optimal point. Since $\mu\neq 0$, it follows from $\partial L/\partial u=0$ with $t>0$ that

$$\cot u=5-2t.$$

This condition may be solved in conjunction with the integrated system equations

$$x_1(t)=2t+t\cos u$$

$$x_2(t)=t\sin u$$

and the equation for the boundary of the target

$$(x_1(t)-5)^2+(x_2(t)-1)^2=1$$

to yield the candidate $u^*=24.56°$ for the optimal control and $t^*=1.406$ for the minimum time to the target. The point where the trajectory intersects the target is given by $x_1(t^*)=4.090$, $x_2(t^*)=0.584$.

In Example 6.4 analytic solutions to the state and cost equations were readily available. When such solutions are not available we have the following result for the case of an unspecified terminal time t^* with the terminal state $x(t^*)$ required to lie in a target of the form (6.23).

Theorem 6.2 Let $u^*\in U$ generate a solution $x^*(t)=\xi(t,u^*)$ to (6.1) defined on $[0,t^*]$. If u^* is a regular local minimal control for the scalar-valued cost $G=z(t,u)$, where $z(t,u)$ is the solution to (6.25) and t^* is an optimal time at which $x(t^*)=\xi(t^*,u^*)\in\Theta$, then there exists an $n\times s$ state perturbation matrix $\Lambda(t, u^*) \triangleq \partial\xi(t, u^*)/\partial u$, satisfying

$$\dot{\Lambda} = \frac{\partial g(x^*(t), u^*)}{\partial x}\Lambda + \frac{\partial g(x^*(t), u^*)}{\partial u} \tag{6.51}$$

$$\Lambda(0, u^*) = [0], \tag{6.52}$$

a $1 \times s$ cost perturbation matrix $\Gamma(t, u^*) \triangleq \partial z(t, u^*)/\partial u$, satisfying

$$\dot{\Gamma} = \frac{\partial \gamma(x^*(t), u^*)}{\partial x}\Lambda + \frac{\partial \gamma(x^*(t), u^*)}{\partial u} \tag{6.53}$$

$$\Gamma(0, u^*) = [0], \tag{6.54}$$

and vectors $\pi \in E^p$, $\mu \in E^q$ such that

$$\gamma(x^*(t^*), u^*) = \pi^T \frac{\partial \theta(x^*(t^*), u^*)}{\partial x} g(x^*(t^*), u^*) \tag{6.55}$$

$$\Gamma(t^*, u^*) = \mu^T \frac{\partial h(u^*)}{\partial u} + \pi^T \left[\frac{\partial \theta(x^*(t^*), u^*)}{\partial x} \Lambda(t^*, u^*) + \frac{\partial \theta(x^*(t^*), u^*)}{\partial u} \right] \tag{6.56}$$

$$\pi^T \theta(x^*(t^*), u^*) = 0 \tag{6.57}$$

$$\theta(x^*(t^*), u^*) \leqq 0 \tag{6.58}$$

$$\pi \leqq 0 \tag{6.59}$$

$$\mu^T h(u^*) \geqq 0 \tag{6.60}$$

$$h(u^*) \geqq 0 \tag{6.61}$$

$$\mu \geqq 0. \tag{6.62}$$

Proof From (6.21) and the definition of a local minimal point we have

$$\gamma[x^*(t^*), u^*]\, \delta t + \Gamma(t^*, u^*) e \geqq 0$$

for all $\delta t \in E^1$ and for all $e \in E^s$ satisfying, from (6.24) and (6.26),

$$\frac{\partial \hat{\theta}(x^*(t^*), u^*)}{\partial x}\left[g(x^*(t^*), u^*)\,\delta t + \Lambda(t^*, u^*)e\right] + \frac{\partial \hat{\theta}(x^*(t^*), u^*)}{\partial u} e \leqq 0$$

$$\frac{\partial \hat{h}(u^*)}{\partial u} e \geqq 0,$$

where $\Lambda(\cdot)$ and $\Gamma(\cdot)$ are the solutions to (6.51) and (6.52) and to (6.53) and (6.54), respectively, and $\hat{\theta}(\cdot)$ and $\hat{h}(\cdot)$ are the active target and control inequality constraints. Thus from Farkas' lemma there exist vectors $\pi \in E^p$ and $\mu \in E^q$, satisfying (6.57)–(6.59) and (6.60)–(6.62), respectively, such that (6.55) and (6.56) hold. ■

The necessary conditions of Theorem 6.2 are similar to the terminality conditions of Vincent and Goh (1972) and the first-order endpoint conditions of Vincent and Brusch (1970). Note that a necessary condition for $x^*(t^*)$ to lie in the interior of Θ is that Theorem 6.2 have a solution with $\pi = 0$.

Theorem 6.2 was developed specifically to handle problems in which the system equations (6.1) cannot be integrated analytically. For such a case Theorem 6.2 is implemented by guessing a value for u^*, then numerically integrating the system equations (6.1) along with the perturbation matrix equations (6.51)–(6.54) until the system trajectory intersects the target at some time t^*. If all of the necessary conditions of Theorem 6.2 are satisfied at this point, then u^* and t^* are candidates for minimizing $G = z(t, u)$.

Note that Theorem 6.1 is a special case of Theorem 6.2. To consider fixed terminal time problems using Theorem 6.2, define $x_{n+1} = t$, add

$$\dot{x}_{n+1} = 1$$

to (6.1), and impose the target conditions

$$\theta_1(\cdot) = x_{n+1}(t^*) - t^* \leqq 0$$

$$\theta_2(\cdot) = t^* - x_{n+1}(t^*) \leqq 0.$$

If the system equations (6.1) and (6.25) can be integrated analytically, yielding explicit solutions $x(t) = \xi(t, u)$ and $G = z(t, u)$, then (6.51)–(6.54) are not required. Instead, the definitions (6.15) and (6.18) for $\Lambda(\cdot)$ and $\Gamma(\cdot)$ are used directly in (6.56).

Example 6.5 Solve Example 6.4 analytically using Theorem 6.2. For the state initially at the origin, the solutions to the cost equation and the state equations are

$$z(t,u)=t$$

$$x_1(t)=\xi_1(t,u)=t(2+\cos u) \tag{6.63}$$

$$x_2(t)=\xi_2(t,u)=t\sin u. \tag{6.64}$$

Thus from (6.15) and (6.18)

$$\Lambda_1(t,u)=\frac{\partial \xi_1}{\partial u}=-t\sin u$$

$$\Lambda_2(t,u)=\frac{\partial \xi_2}{\partial u}=t\cos u$$

$$\Gamma(t,u)=\frac{\partial z}{\partial u}=0.$$

Then with target (6.49) and u unconstrained, (6.55) and (6.56) yield

$$1=\pi\left[2(x_1(t)-5)(2+\cos u)+2(x_2(t)-1)\sin u\right] \tag{6.65}$$

$$0=\pi\left[-2(x_1(t)-5)t\sin u+2(x_2(t)-1)t\cos u\right], \tag{6.66}$$

where $\pi \leqq 0$. It follows from (6.65) that $\pi \neq 0$, so that (6.66) yields

$$\tan u=\frac{x_2(t)-1}{x_1(t)-5}.$$

It follows from (6.63) and (6.64) that

$$\frac{x_2(t)}{x_1(t)-2}=\tan u.$$

Hence $x_1(t)$ and $x_2(t)$ may be determined from

$$\frac{x_2(t)}{x_1(t)-2}=\frac{x_2(t)-1}{x_1(t)-5}$$

and

$$(x_1(t)-5)^2+(x_2(t)-1)^2=1,$$

which yields the same results as before.

In the following example analytic solutions are also easily obtained for the state and cost equations. However, in order to illustrate Theorem 6.2 more fully, we do not make use of these solutions at the outset.

Example 6.6 Determine the constant control u that will transfer the system

$$\dot{x}_1 = x_2 \tag{6.67}$$

$$\dot{x}_2 = u \tag{6.68}$$

from the origin to the line $x_1 = 10$ in minimum time, subject to the constraint $|u| \leqq 1$.

The target set and control constraints yield the following inequalities:

$$\theta_1(\cdot) = x_1 - 10 \leqq 0 \tag{6.69}$$

$$\theta_2(\cdot) = 10 - x_1 \leqq 0 \tag{6.70}$$

$$h_1(\cdot) = 1 - u \geqq 0 \tag{6.71}$$

$$h_2(\cdot) = u + 1 \geqq 0. \tag{6.72}$$

The rate of change of cost is simply

$$\dot{G} = 1. \tag{6.73}$$

From

$$\frac{\partial g}{\partial x} = \begin{bmatrix} 0 & 1 \\ 0 & 0 \end{bmatrix}$$

and

$$\frac{\partial g}{\partial u} = \begin{bmatrix} 0 \\ 1 \end{bmatrix}$$

the state perturbation equations are

$$\dot{\Lambda}_1 = \Lambda_2 \qquad \Lambda_1(0, u) = 0 \tag{6.74}$$

$$\dot{\Lambda}_2 = 1 \qquad \Lambda_2(0, u) = 0. \tag{6.75}$$

From

$$\frac{\partial \gamma}{\partial x} = [0,0]$$

and

$$\frac{\partial \gamma}{\partial u} = 0$$

the scalar cost perturbation equations are

$$\dot{\Gamma} = 0 \qquad \Gamma(0, u) = 0. \tag{6.76}$$

From (6.76) we have

$$\Gamma(t, u) \equiv 0. \tag{6.77}$$

Integrating (6.67), (6.68), (6.74), and (6.75) we have

$$x_1(t) = \xi_1(t, u) = \tfrac{1}{2} u t^2 \tag{6.78}$$

$$x_2(t) = \xi_2(t, u) = ut \tag{6.79}$$

$$\Lambda_1(t, u) = \tfrac{1}{2} t^2 \tag{6.80}$$

$$\Lambda_2(t, u) = t. \tag{6.81}$$

At the final time $t = t^*$ the necessary conditions (6.55) and (6.56) yield

$$1 = [\pi_1 - \pi_2] x_2(t) \tag{6.82}$$

$$0 = -\mu_1 + \mu_2 + [\pi_1 - \pi_2] \Lambda_1(t, u). \tag{6.83}$$

Using (6.79)–(6.81) in (6.82) and (6.83), we have

$$1 = [\pi_1 - \pi_2] ut \tag{6.84}$$

$$0 = -\mu_1 + \mu_2 + \tfrac{1}{2} [\pi_1 - \pi_2] t^2, \tag{6.85}$$

which yield

$$t^* = 2u^*(\mu_1 - \mu_2) \tag{6.86}$$

where $\mu_1 \geqq 0$, $\mu_2 \geqq 0$, $\pi_1 \leqq 0$, $\pi_2 \leqq 0$ and

$$\mu_1(1-u) = 0$$

$$\mu_2(u+1) = 0.$$

For $t^* > 0$ it follows from (6.86) that μ_1 and μ_2 are not both zero. Hence one of the control constraints is active, that is, $u^* = \pm 1$. From (6.69), (6.70), and (6.78) we have

$$x_1(t^*) = 10 = \tfrac{1}{2}u^* t^{*2} > 0.$$

Therefore we conclude

$$u^* = 1.$$

Sometimes it may be convenient to convert algebraic equations into differential equations and employ Theorem 6.2 even though the problem could be formulated otherwise.

Example 6.7 (Vincent and Mason, 1969) Three straight lines are to be drawn through three noncolinear points in a plane (the initial points) such that the lines intersect at a common fourth point. The problem is to determine the location of the fourth point so that the sum of the distances from the three initial points to the fourth point is a minimum.

Without loss of generality we may locate one of the initial points at the origin of an axis system with the other two initial points located in the positive orthant, as shown in Figure 6.2. The fourth point is denoted by the coordinates y_1, y_2. Designate the slope of each line by u_i; thus

$$\tan u_i = \frac{y_2 - y_2^i(0)}{y_1 - y_1^i(0)} \qquad i = 1,2,3$$

where $y_1^i(0)$ and $y_2^i(0)$, $i = 1,2,3$ are the coordinates locating points 1, 2, and 3. Thus

$$y_2 - y_2^i(0) = \left[y_1 - y_1^i(0) \right] \tan u_i . \tag{6.87}$$

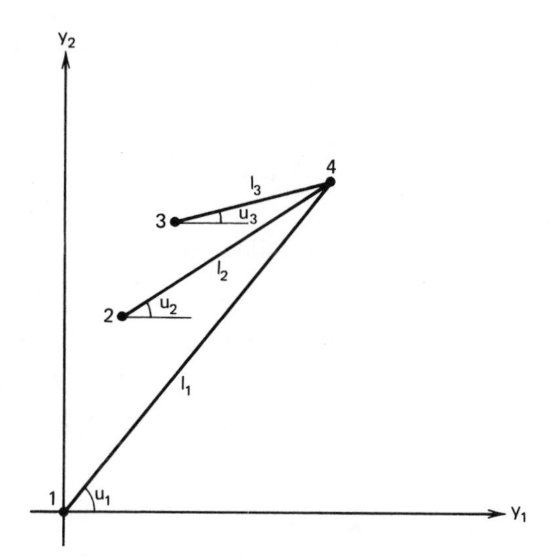

Figure 6.2. Geometry for Example 6.7.

Now define

$$x_i \triangleq y_2 - y_2^i(0)$$

$$t_i \triangleq y_1 - y_1^i(0) \tag{6.88}$$

so that (6.87) may be written as

$$x_i = t_i \tan u_i.$$

Thus

$$\frac{dx_i}{dt_i} = \tan u_i.$$

Note, however, that $dt_i \equiv dy_1$ for $i=1,2,3$. Define $dy_1 \triangleq dt$ for convenient use of the Theorem 6.2. Thus from (6.88) we have the system defined by

$$\dot{x}_i = \tan u_i \qquad i=1,2,3 \tag{6.89}$$

and the problem is to determine u_1, u_2, and u_3 to minimize the sum of the lengths of the line segments from the three initial points to the fourth point. The sum of lengths of the three segments is given by

$$G=\sum_{i=1}^{3} l_i=\sum_{i=1}^{3}\left[y_1-y_1^i(0)\right]\sec u_i. \tag{6.90}$$

Using the previous definitions, it follows that

$$\dot{G}=\sum_{i=1}^{3}\sec u_i. \tag{6.91}$$

Each segment must intersect at the common fourth point. At the final time

$$x_1=x_2=x_3$$

or equivalently

$$\begin{aligned}\theta_1&=x_1-x_2\leqq 0\\ \theta_2&=x_1-x_3\leqq 0\\ \theta_3&=x_2-x_1\leqq 0\\ \theta_4&=x_3-x_1\leqq 0.\end{aligned}$$

For this problem the solution to (6.89) is given by (6.87) and the solution (6.91) is given by (6.90). The necessary condition (6.55) yields

$$\begin{aligned}\sec u_1+\sec u_2+\sec u_3=&(\pi_1-\pi_3)(\tan u_1-\tan u_2)\\ &+(\pi_2-\pi_4)(\tan u_1-\tan u_3),\end{aligned}$$

where $\pi_j\leqq 0$, $j=1,\ldots,4$. Utilizing (6.15), (6.18), $y_1^1(0)=y_2^1(0)=0$, and $t=y_1$, the necessary conditions (6.56) yield

$$\begin{aligned}t\sec u_1\tan u_1&=(\pi_1-\pi_3)t\sec^2 u_1+(\pi_2-\pi_4)t\sec^2 u_1\\ \left[t-y_1^2(0)\right]\sec u_2\tan u_2&=(\pi_3-\pi_1)\left[t-y_1^2(0)\right]\sec^2 u_2\\ \left[t-y_1^3(0)\right]\sec u_3\tan u_3&=(\pi_4-\pi_2)\left[t-y_1^3(0)\right]\sec^2 u_3.\end{aligned}$$

It follows from the above three equations that

$$\pi_3 - \pi_1 = \sin u_2,$$

$$\pi_4 - \pi_2 = \sin u_3,$$

and

$$\sin u_1 + \sin u_2 + \sin u_3 = 0. \tag{6.92}$$

Using these results with the first necessary condition yields

$$\frac{1}{\cos u_1} + \frac{1}{\cos u_2} + \frac{1}{\cos u_3} + \frac{\sin u_1 \sin u_2}{\cos u_1} + \frac{\sin u_3 \sin u_1}{\cos u_1} - \frac{\sin^2 u_2}{\cos u_2} - \frac{\sin^2 u_3}{\cos u_3} = 0,$$

which reduces to

$$\cos u_1 + \cos u_2 + \cos u_3 = 0. \tag{6.93}$$

Equations (6.92) and 96.93) are satisfied by

$$u_2 = u_1 + 120° \tag{6.94}$$

$$u_3 = u_1 - 120°. \tag{6.95}$$

Thus any optimal solution has a simple geometric relation between the slopes. If we choose a value for u_1 ($90° < u_1 < -90°$) and a final "time" t, then the common point will always be at $(t, t \tan u_1)$. We can then lay off the slope of the other two lines and locate the other two points by moving a distance $y_1 = t$ in the appropriate direction.

6.5 PARAMETRIC DIFFERENTIAL GAMES

In this section we consider r-player games with a dynamic state $x \in E^n$ governed by (6.1) with initial state $x(0)$ specified. The cost vector $G = (G_1, \ldots, G_r)$ is also dynamic, with its rate of change governed by (6.5). The composite control vector $u = (u^1, \ldots, u^r) \in E^s = E^{s_i} \times \cdots \times E^{s_r}$ is a constant and is subject to the constraints $u \in U$, where $U \subseteq E^s$ is a completely regular set given by

$$U=\{u\in E^s|h(u)\geqq 0\}$$

where $h(\cdot): E^s \rightarrow E^q$ is C^1 in a ball about a solution point u (to be specified).

The basic objective of each Player i, $i=1,\ldots,r$, is to minimize his cost $G_i=z_i(t,u)$ evaluated at a final time $t=t_f$ at which $x(t)=\xi(t,u)\in\Theta$ with a regular target set Θ given by (6.23) where $\xi(t,u)$ is the solution to (6.1), generated by $u\in U$ and satisfying $\xi(0,u)=x(0)$, and $z(t,u)=[z_1(t,u),\ldots,z_r(t,u)]^T$ is the corresponding solution to (6.5) with $z(0,u)=[0]$.

The solution concepts that we consider are Nash equilibrium, min-max, and Pareto-minimal solutions.

6.5.1 Nash Equilibrium Solutions

Let $\hat{u}=(\hat{u}^i,\hat{v})\in U$ be a completely regular point, where u^i is the control for Player i and v is the composite control for the remaining players. Let $\hat{t}$ denote the final time, let $\hat{t}+\alpha\delta t$ be a perturbation in the final time, and let $\hat{u}^i+\alpha\delta u^i(\alpha)$ be a perturbation in the control for Player i (with all other controls held fixed) where $\delta u^i(\alpha)\rightarrow e^i\in T_i$ and T_i is the tangent cone to $U_i \triangleq \{u^i\in E^{s_i}|h(u^i,\hat{v})\geqq 0\}$, given by

$$T_i=\left\{e^i\in E^{s_i}\,\middle|\,\frac{\partial\hat{h}(\hat{u})}{\partial u^i}e^i\geqq 0\right\},$$

where $\hat{h}(\cdot)$ denotes the active control inequalities at $\hat{u}\in U$.

At a completely regular local Nash equilibrium point $\hat{u}\in U$ we have, from (6.11) as $\alpha\rightarrow 0$, for each $i=1,\ldots,r$

$$\gamma_i\big(\xi(\hat{t},\hat{u}),\hat{u}\big)\,\delta t+\frac{\partial z_i(\hat{t},\hat{u})}{\partial u^i}e^i\geqq 0 \tag{6.96}$$

for all $\delta t\in E^1$ and $e^i\in E^{s_i}$ such that

$$\frac{\partial\hat{\theta}\left[\xi(\hat{t},\hat{u}),\hat{u}\right]}{\partial x}\left[g\left[\xi(\hat{t},\hat{u}),\hat{u}\right]\delta t+\frac{\partial\xi(\hat{t},\hat{u})}{\partial u^i}e^i\right]+\frac{\partial\hat{\theta}\left[\xi(\hat{t},\hat{u}),\hat{u}\right]}{\partial u^i}e^i\leqq 0, \tag{6.97}$$

$$\frac{\partial\hat{h}(\hat{u})}{\partial u^i}e^i\geqq 0, \tag{6.98}$$

where $\hat{\theta}(\cdot)$ denotes the active target inequalities at $\xi(\hat{t},\hat{u})\in\Theta$, and $\hat{h}(\cdot)$ denotes the active control inequalities at $\hat{u}$. Condition (6.97) follows from the requirement $\xi(\hat{t}+\alpha\,\delta t,\,\hat{u}^i+\alpha\,\delta u^i(\alpha),\,\hat{v})\in\Theta$.

Defining the $1\times s_i$ *cost perturbation matrix for Player i* as

$$T^i(t,u)\triangleq\frac{\partial z_i(t,u)}{\partial u^i}\tag{6.99}$$

and the $n\times s_i$ *state perturbation matrix for Player i* as

$$\Lambda^i(t,u)\triangleq\frac{\partial \xi(t,u)}{\partial u^i},\tag{6.100}$$

we may write (6.96) and (6.97) as

$$\gamma_i\big[\xi(\hat{t},\hat{u}),\hat{u}\big]\,\delta t+\Gamma^i(\hat{t},\hat{u})e^i\geqq 0\tag{6.101}$$

and

$$\frac{\partial\hat{\theta}\,\big[\xi(\hat{t},\hat{u}),\hat{u}\big]}{\partial x}\big[g\big[\xi(\hat{t},\hat{u}),\hat{u}\big]\,\delta t+\Lambda^i(\hat{t},\hat{u})e^i\big]+\frac{\partial\hat{\theta}\,\big[\xi(\hat{t},\hat{u}),\hat{u}\big]}{\partial u^i}e^i\leqq 0,\tag{6.102}$$

where, from Leibniz' rule and the solutions $\xi(t,u)$ and $z(t,u)$ to (6.1) and (6.5)

$$\dot{\Lambda}^i=\frac{\partial g\big[\xi(t,\hat{u}),\hat{u}\big]}{\partial x}\Lambda^i+\frac{\partial g\big[\xi(t,\hat{u}),\hat{u}\big]}{\partial u^i}\tag{6.103}$$

$$\Lambda^i(0,\hat{u})=[0]\tag{6.104}$$

$$\dot{\Gamma}^i=\frac{\partial \gamma_i\big[\xi(t,\hat{u}),\hat{u}\big]}{\partial x}\Lambda^i+\frac{\partial \gamma_i\big[\xi(t,\hat{u}),\hat{u}\big]}{\partial u^i}\tag{6.105}$$

$$\Gamma^i(0,\hat{u})=[0].\tag{6.106}$$

Thus we have the following theorem.

Theorem 6.3 If $\hat{u}\in U$ is a completely regular local Nash equilibrium point for the cost vector $G=z(t,u)$ at time $t=\hat{t}$, with corresponding state $\hat{x}(t)=\xi(t,\hat{u})$ satisfying (6.1) with $\hat{x}(\hat{t})\in\Theta$, then for each $i=1,\ldots,r$ there exists a $1\times s_i$ cost perturbation matrix $\Gamma^i(t,\hat{u})\triangleq\partial z_i(t,\hat{u})/\partial u^i$, satisfying (6.105) and (6.106), and an $n \times s_i$ state perturbation matrix $\Lambda^i(t, \hat{u}) \triangleq \partial\xi(t, \hat{u})/\partial u^i$, satisfying (6.103) and (6.104), and there exist vectors $\pi(i) \in E^p$ and $\mu(i) \in E^q$ such that

$$\gamma_i[\hat{x}(\hat{t}), \hat{u}] = \pi^T(i) \frac{\partial\theta[\hat{x}(\hat{t}), \hat{u}]}{\partial x} g(\hat{x}(\hat{t}), \hat{u}) \tag{6.107}$$

$$\Gamma^i(\hat{t}, \hat{u}) = \mu^T(i) \frac{\partial h(\hat{u})}{\partial u^i} + \pi^T(i) \left[\frac{\partial\theta(\hat{x}(\hat{t}), \hat{u})}{\partial x} \Lambda^i(\hat{t}, \hat{u}) + \frac{\partial\theta(\hat{x}(\hat{t}), \hat{u})}{\partial u^i} \right] \tag{6.108}$$

$$\pi^T(i)\theta(\hat{x}(\hat{t}), \hat{u}) = 0 \tag{6.109}$$

$$\theta(\hat{x}(\hat{t}), \hat{u}) \leqq 0 \tag{6.110}$$

$$\pi(i) \leqq 0 \tag{6.111}$$

$$\mu^T(i) h(\hat{u}) = 0 \tag{6.112}$$

$$h(\hat{u}) \geqq 0 \tag{6.113}$$

$$\mu(i) \geqq 0. \tag{6.114}$$

For a specified terminal time $\hat{t}$ with no constraints on the final state, condition (6.107) does not apply and the remaining conditions (6.108)–(6.114) hold with $\pi(i) = 0, i = 1, \ldots, r$.

Proof The theorem follows from Farkas' lemma applied to (6.96)–(6.98) using (6.99)–(6.106). ■

Note that the Nash equilibrium necessary conditions of Theorem 6.3 constitute a two-point boundary value problem involving $n + (n+1)s$ differential equations, where $s = s_1 + \cdots + s_r$ is the dimension of the composite control vector.

Example 6.8 Consider two genotypes of a particular species of annual plant growing together in a plot of land and competing for a limited nutrient resource. For two genotypes a nondimensional Lotka-Volterra model (Vincent, 1978b) is

$$\dot{x}_1 = u_1 x_1(1 - u_1 x_1 - u_2 x_2) \tag{6.115}$$

$$\dot{x}_2 = u_2 x_2(1 - u_1 x_1 - u_2 x_2) \tag{6.116}$$

where $x_i \in E^1$, $x_i > 0$ is the biomass of genotype i and $u_i \in E^1$, $u_i > 0$, is an adaptive (i.e., control) parameter associated with genotype i.

We consider the case where each genotype chooses its adaptive parameter u_i to maximize its "fitness" at the end of a specified growing season, that is, to maximize $x_i(\hat{t}) - x_i(0)$ given $\hat{t}$. Then the cost vector $G = (G_1, G_2)$ satisfies

$$\dot{G}_1 = -\dot{x}_1 \tag{6.117}$$

$$\dot{G}_2 = -\dot{x}_2 \tag{6.118}$$

and the fixed terminal time target is given by

$$\theta_1(\cdot) = x_3(\hat{t}) - \hat{t} \leqq 0 \tag{6.119}$$

$$\theta_2(\cdot) = \hat{t} - x_3(\hat{t}) \leqq 0 \tag{6.120}$$

where time $t = x_3$ is appended to the state equations via

$$\dot{x}_3 = 1 \tag{6.121}$$

with $x_3(0) = 0$. Then the state perturbation equations (6.103) are

$$\dot{\Lambda}_1^1 = (u_1 - 2u_1^2 x_1 - u_1 u_2 x_2)\Lambda_1^1 - u_1 u_2 x_1 \Lambda_2^1 + x_1 - 2u_1 x_1^2 - u_2 x_1 x_2 \tag{6.122}$$

$$\dot{\Lambda}_2^1 = -u_1 u_2 x_2 \Lambda_1^1 + (u_2 - 2u_2^2 x_2 - u_2 u_1 x_1)\Lambda_2^1 - u_2 x_1 x_2 \tag{6.123}$$

$$\dot{\Lambda}_3^1 = 0 \tag{6.124}$$

$$\dot{\Lambda}_1^2 = (u_1 - 2u_1^2 x_1 - u_1 u_2 x_2)\Lambda_1^2 - u_1 u_2 x_1 \Lambda_2^2 - u_1 x_2 x_1 \tag{6.125}$$

$$\dot{\Lambda}_2^2 = -u_1 u_2 x_2 \Lambda_1^2 + (u_2 - 2u_2^2 x_2 - u_2 u_1 x_1)\Lambda_2^2 + x_2 - 2u_2 x_2^2 - u_1 x_2 x_1 \tag{6.126}$$

$$\dot{\Lambda}_3^2 = 0 \tag{6.127}$$

with initial conditions $\Lambda^i_j(0, u)=0$, $i=1,2$; $j=1,2,3$. The cost perturbation equations (6.105) are

$$\dot{\Gamma}^1=(-u_1+2u_1^2x_1+u_1u_2x_2)\Lambda_1^1+u_1u_2x_1\Lambda_2^1-x_1+2u_1x_1^2+u_2x_1x_2 \tag{6.128}$$

$$\dot{\Gamma}^2=u_1u_2x_2\Lambda_1^2+(-u_2+2u_2^2x_2+u_1u_2x_1)\Lambda_2^2-x_2+2u_2x_2^2+u_1x_1x_2 \tag{6.129}$$

with initial conditions $\Gamma^i(0, u)=0$, $i=1,2$.

Since the constraints $u_i>0$, $i=1,2$, define open constraints sets, the terminal conditions (6.107) and (6.108) are

$$-u_1x_1(\hat{t})[1-u_1x_1(\hat{t})-u_2x_2(\hat{t})]=\pi_1(1)-\pi_2(1) \tag{6.130}$$

$$-u_2x_2(\hat{t})[1-u_1x_1(\hat{t})-u_2x_2(\hat{t})]=\pi_1(2)-\pi_2(2) \tag{6.131}$$

$$\Gamma^1(\hat{t}, u)=\Lambda_3^1(\hat{t}, u)[\pi_1(1)-\pi_2(1)] \tag{6.132}$$

$$\Gamma^2(\hat{t}, u)=\Lambda_3^2(\hat{t}, u)[\pi_1(2)-\pi_2(2)], \tag{6.133}$$

where $\pi(1)\leqq 0, \pi(2)\leqq 0$. From (6.124) and (6.127) with zero initial conditions we have $\Lambda_3^1(t, u)=\Lambda_3^2(t, u)\equiv 0$. Thus (6.132) and (6.133) become

$$\Gamma^1(\hat{t}, u)=\Gamma^2(\hat{t}, u)=0 \tag{6.134}$$

and (6.130) and (6.131) yield no additional useful information.

Thus Nash equilibrium controls $\hat{u}_1$ and $\hat{u}_2$, generally determined in an iterative fashion, are such that (6.134) is satisfied when the state equations (6.115) and (6.116) are integrated to time $\hat{t}$ starting from an initial point $x(0)$ and the state perturbation equations (6.122), (6.123), (6.125), and (6.126) and the cost perturbation equations (6.128) and (6.129) are integrated to time $\hat{t}$ starting from zero initial conditions.

It is of interest to compare some results obtained this way with the results obtained in Example 6.2. Note that the model given by (6.115) and (6.116) reduces to the single species model (6.28) if we add (6.115) and (6.116) and let $u_1=u_2=u$ and $x_1+x_2=x$. This might lead us to believe that the Nash solution should be the same as given by (6.31).

This is not the case. For example, (6.31) yields $u^*t^*=2.93$ if $t^*/x(0)=20.66$, whereas the numerical procedure just outlined yields $u_1^*t^*=u_2^*t^*=4.25$ if $t^*/x_1(0)=t^*/x_2(0)=20.66$. Thus for a fixed growing season t^* and initial conditions $x_1(0)$ and $x_2(0)$, the adaptive parameter is seen to increase in a competitive situation. This results in a smaller total yield for the system.

6.5.2 Min-Max Solutions

The basis for the Nash equilibrium solution is a presumption that each player acts independently and seeks to minimize his own final cost. The min-max solution for a particular player (Player 1) assumes that the other players act collectively (Player 2) with the objective of maximizing the cost to Player 1.

For a min-max solution we consider a two-player zero-sum game with control $u=(u^1,u^2)\in E^{s_1}\times E^{s_2}=E^s$, $s=s_1+s_2$. Player 1 controls u^1 and seeks to minimize a scalar-valued cost

$$G=\int_0^t \gamma(x,u)\,dt$$

at some final time $t=t^*$. Player 2 controls u^2 and seeks to maximize G. The state at time $t\in[0,t^*]$ is given by $x(t)=\xi(t,u)$ where $\xi(\cdot): E^1\times E^s\to E^n$ is the solution to (6.1) starting at some specified initial state. The cost to Player 1 at time t is given by $G=z(t,u)$ where $z(\cdot): E^1\times E^s\to E^1$ is the solution to

$$\dot{G}=\gamma[\xi(t,u),u]$$

with $z(0,u)=0$.

The control is restricted to $u\in U$ where U is of the form (6.2). The final state is restricted to lie in a regular target set $\Theta\subseteq E^n$ of the form (6.23). The final time is otherwise unspecified, although fixed final time problems may be treated by augmenting the state equations, as discussed in Section 6.4.

Since the Nash equilibrium and min-max solutions (if they exist) are identical in a two-player zero-sum game, we have the following theorem.

Theorem 6.4 If $u^*=(u^{1*},u^{2*})$ is a completely regular local min-max solution for Player 1, with state $x^*(t)=\xi(t,u^*)$ and final time t^* such that $x^*(t^*)\in\Theta$, then for each $i=1,2$ there exists an $n\times s_i$ state perturbation matrix $\Lambda^i(t, u^*) \triangleq \partial\xi(t, u^*)/\partial u^i$, satisfying

$$\dot{\Lambda}^i = \frac{\partial g(x^*(t), u^*)}{\partial x}\Lambda^i + \frac{\partial g(x^*(t), u^*)}{\partial u^i} \tag{6.135}$$

$$\Lambda^i(0, u^*) = 0, \tag{6.136}$$

a $1 \times s_i$ cost perturbation matrix $\Gamma^i(t, u^*) \triangleq \partial z(t, u^*)/\partial u^i$, satisfying

$$\dot{\Gamma}^i = \frac{\partial \gamma(x^*(t), u^*)}{\partial x}\Lambda^i + \frac{\partial \gamma(x^*(t), u^*)}{\partial u^i} \tag{6.137}$$

$$\Gamma^i(0, u^*) = 0, \tag{6.138}$$

and vectors $\mu(i) \in E^q$ and $\pi(i) \in E^p$ such that the following conditions hold for $i = 1, 2$:

$$\gamma(x^*(t^*), u^*) = \pi^T(i)\frac{\partial \theta(x^*(t^*), u^*)}{\partial x} g(x^*(t^*), u^*) \tag{6.139}$$

$$\Gamma^i(t^*, u^*) = \mu^T(i)\frac{\partial h(u^*)}{\partial u^i} + \pi^T(i)\left[\frac{\partial \theta(x^*(t^*), u^*)}{\partial x}\Lambda^i(t^*, u^*) + \frac{\partial \theta(x^*(t^*), u^*)}{\partial u^i}\right] \tag{6.140}$$

and

$$h(u^*) \geqq 0 \tag{6.141}$$

$$\mu^T(i) h(u^*) = 0 \qquad i = 1, 2 \tag{6.142}$$

$$\mu(1) \geqq 0 \tag{6.143}$$

$$\mu(2) \leqq 0 \tag{6.144}$$

$$\theta(x^*(t^*), u^*) \leqq 0 \tag{6.145}$$

$$\pi^T(i)\theta(x^*(t^*), u^*) = 0 \qquad i = 1, 2 \tag{6.146}$$

$$\pi(1) \leqq 0 \tag{6.147}$$

$$\pi(2) \geqq 0. \tag{6.148}$$

Proof The theorem follows directly from Theorem 6.3 by use of a two dimensional cost vector $(G, -G)$. ■

For the special case of a specified terminal time t^* with no constraints on the final state, the min-max necessary conditions (6.139)–(6.148) simplify to

$$\Gamma^1(t^*, u^*) = \mu^T(1)\frac{\partial h(u^*)}{\partial u^1} \tag{6.149}$$

$$\Gamma^2(t^*, u^*) = \mu^T(2)\frac{\partial h(u^*)}{\partial u^2} \tag{6.150}$$

$$h(u^*) \geqq 0 \tag{6.151}$$

$$\mu^T(1)h(u^*) = 0 \tag{6.152}$$

$$\mu^T(2)h(u^*) = 0 \tag{6.153}$$

$$\mu(1) \geqq 0 \tag{6.154}$$

$$\mu(2) \leqq 0. \tag{6.155}$$

Example 6.9 Consider a pursuer (Player 1) and an evader (Player 2) moving in a plane with constant speeds V_P and V_E and constant headings u_1 and u_2, respectively. Let $x = (x_1, x_2)$ denote the position of the evader with respect to the pursuer. Then

$$\dot{x}_1 = V_E \cos u_2 - V_P \cos u_1 = g_1(\cdot)$$

$$\dot{x}_2 = V_E \sin u_2 - V_P \sin u_1 = g_2(\cdot).$$

For a specified terminal time t^* the pursuer seeks to minimize and the evader seeks to maximize

$$G = \tfrac{1}{2}\left[r^2(t^*) - r^2(0)\right] = \int_0^{t^*} r\dot{r}\, dt$$

where $r = \sqrt{x_1^2 + x_2^2}$ is the distance between the pursuer and the evader. Thus

$$\dot{G} = x^T g = \gamma(\cdot)$$

and we have a particular case of the fixed-time problem with

$$\dot{G} = \frac{\partial V(x)}{\partial x} g(u) = \gamma(x, u)$$

$$\dot{x} = g(u).$$

For our pursuit problem

$$V(x) = \tfrac{1}{2} x^T A x$$

where A is the 2×2 identity matrix. Since $\partial g / \partial x = 0$ (the velocity vector $\dot{x}$ is a constant), the 2×2 composite state perturbation matrix $\Lambda(t,u) \triangleq \partial \xi(t,u)/\partial u = [\Lambda^1, \Lambda^2]$ is the solution to

$$\dot{\Lambda} = \frac{\partial g}{\partial u}$$

$$\Lambda(0, u) = [0],$$

so that

$$\Lambda(t, u) = \frac{\partial g}{\partial u} t,$$

and the state is given by

$$x(t) = x(0) + gt.$$

The 1×2 composite cost perturbation matrix $\Gamma(t,u) \triangleq \partial z(t,u)/\partial u = [\Gamma^1, \Gamma^2]$ is the solution to

$$\begin{aligned} \dot{\Gamma} &= \frac{\partial \gamma}{\partial x} \Lambda + \frac{\partial \gamma}{\partial u} \\ &= \left[g^T \frac{\partial^2 V}{\partial x^2} t + \frac{\partial V}{\partial x} \right] \frac{\partial g}{\partial u} \\ &= \frac{d}{dt} \left[t \frac{\partial V}{\partial x} \right] \frac{\partial g}{\partial u}, \end{aligned}$$

satisfying the initial condition $\Gamma(0, u)=[0]$. Therefore

$$\Gamma(t, u)=t\frac{\partial V}{\partial x}\frac{\partial g}{\partial u}.$$

For *any* specified terminal time $t^*>0$ and initial state $x(0)$, with no constraints on the control u or the final state $x(t^*)$, the necessary conditions (6.149)–(6.155) reduce to $\Gamma(t^*, u)=[0]$, yielding

$$\frac{\partial V}{\partial x}\frac{\partial g}{\partial u}=0.$$

Note that these are the same as the necessary conditions for a minimum, maximum, or min-max of the *instantaneous* rate of change of the cost $G(t, u)=V[x(t)]$. Thus the original *integral* cost problem reduces to a *point* problem. For $\partial V/\partial x\neq 0$ each of the vectors $\partial g/\partial u_1$, $\partial g/\partial u_2$ must be orthogonal to $\partial V/\partial x$. For our pursuit-evasion problem we have $\partial V/\partial x=x^T$. Thus

$$[0,0]=[x_1, x_2]\begin{bmatrix} V_P\sin u_1^* & -V_E\sin u_2^* \\ -V_P\cos u_1^* & V_E\cos u_2^* \end{bmatrix}.$$

For $x\neq 0$ we have

$$0=\left|\frac{\partial g}{\partial u}\right|=V_E V_P\sin(u_1^*-u_2^*)$$

and

$$0=[x_1\sin u_1^*-x_2\cos u_1^*]V_P.$$

Thus (for min-max)

$$u_1^*=u_2^*=\tan^{-1}\left[\frac{x_2}{x_1}\right]=\tan^{-1}\left[\frac{x_2(0)}{x_1(0)}\right].$$

The optimal strategy $u_2[x(0)]$ for the evader is to head away from the pursuer along the initial line, from the pursuer through the evader, and the optimal strategy $u_1[x(0)]$ for the pursuer is to follow the evader in the same direction (tail chase).

6.5.3 Pareto-Minimal Solutions

Consider the game with vector-valued cost at time t given by $G=[G_1,\ldots,G_r]^T=z(t,u)$ where $z(\cdot)$ is the solution to (6.5), with $z(0,u)=0$, corresponding to the composite control $u=(u^1,\ldots,u^r)\in U$ and state $x(t)=\xi(t,u)$, where U is given by (6.2) and $\xi(\cdot)$ is the solution to (6.1) with specified initial state $x(0)=\xi(0,u)$. Let t° denote the final time, which satisfies $x(t^\circ)\in\Theta$, where Θ is of the form (6.23).

With cooperation possible among the players we consider the Pareto-minimal solution corresponding to the case where the players cooperate so long as it is not to their own disadvantage.

Let u° denote a regular local Pareto-minimal control with final time t°. Let $t^\circ+\alpha\,\delta t$ be a perturbation in the final time, and let $u^\circ+\alpha\,\delta u(\alpha)\in U$ be a perturbation in the control, with $\delta u(\alpha)\to e\in T_U$ as $\alpha\to 0$, where T_U is the tangent cone to U at u°.

In terms of δG, given by (6.21), the basic necessary conditions at the Pareto-minimal point u°, with corresponding state $x^\circ(t)=\xi(t,u^\circ)$, are that there do *not* exist perturbations $\delta t\in E^1$, $e\in T_U$ such that $\delta G\leqslant 0$ at the final time t°. Therefore, from Chapter 3, there exists a vector $\eta\in E^r$, with $\eta\geqslant 0$, such that

$$\eta^T\left[\gamma(\xi(t^\circ,u^\circ),u^\circ)\,\delta t+\frac{\partial z(t^\circ,u^\circ)}{\partial u}e\right]\geqq 0 \tag{6.156}$$

for all $\delta t\in E^1$ and for all $e\in E^s$ such that

$$\frac{\partial \hat{h}(u^\circ)}{\partial u}\geqq 0 \tag{6.157}$$

$$\frac{\partial\hat{\theta}\,(\xi(t^\circ,u^\circ),u^\circ)}{\partial x}\left[g(\xi(t^\circ,u^\circ),u^\circ)\,\delta t+\Lambda(t^\circ,u^\circ)e\right]+\frac{\partial\hat{\theta}\,(\xi(t^\circ,u^\circ),u^\circ)}{\partial u}e\leqq 0. \tag{6.158}$$

where $\hat{h}(\cdot)$ and $\hat{\theta}(\cdot)$ denote the active control and target constraints, respectively. Then we have the following theorem.

Theorem 6.5 If $u^\circ\in U$ is a regular local Pareto-minimal solution, with state $x^\circ(t)=\xi(t,u^\circ)$ and final time t° such that $x^\circ(t^\circ)\in\Theta$, then there exists an $n \times s$ state perturbation matrix $\Lambda(t, u^\circ) \triangleq \partial\xi(t, u^\circ)/\partial u^\circ$, satisfying

$$\dot{\Lambda} = \frac{\partial g(x^\circ(t), u^\circ)}{\partial x}\Lambda + \frac{\partial g(x^\circ(t), u^\circ)}{\partial u} \tag{6.159}$$

$$\Lambda(0, u^\circ) = [0], \tag{6.160}$$

there exists a vector $\eta \in E^r$ and a $1 \times s$ cost perturbation matrix $\Gamma(t, u^\circ) \triangleq \eta^T \partial z(t^\circ, u^\circ)/\partial u$ satisfying

$$\dot{\Gamma} = \eta^T\left[\frac{\partial \gamma(x^\circ(t), u^\circ)}{\partial x}\Lambda + \frac{\partial \gamma(x^\circ(t), u^\circ)}{\partial u}\right] \tag{6.161}$$

$$\Gamma(0, u^\circ) = [0], \tag{6.162}$$

and there exist vectors $\mu \in E^q$ and $\pi \in E^p$ such that

$$\eta^T \gamma(x^\circ(t^\circ), u^\circ) = \pi^T \frac{\partial \theta(x^\circ(t^\circ), u^\circ)}{\partial x} g(x^\circ(t^\circ), u^\circ) \tag{6.163}$$

$$\Gamma(t^\circ, u^\circ) = \mu^T \frac{\partial h(u^\circ)}{\partial u} + \pi^T\left[\frac{\partial \theta(x^\circ(t^\circ), u^\circ)}{\partial x}\Lambda(t^\circ, u^\circ) + \frac{\partial \theta(x^\circ(t^\circ), u^\circ)}{\partial u}\right] \tag{6.164}$$

$$h(u^\circ) \geqq 0 \tag{6.165}$$

$$\mu^T h(u^\circ) = 0 \tag{6.166}$$

$$\mu \geqq 0 \tag{6.167}$$

$$\theta(x^\circ(t^\circ), u^\circ) \leqq 0 \tag{6.168}$$

$$\pi^T \theta(x^\circ(t^\circ), u^\circ) = 0 \tag{6.169}$$

$$\pi \leqq 0 \tag{6.170}$$

$$\eta \geqslant 0. \tag{6.171}$$

Proof The theorem follows from Farkas' lemma applied to (6.156)–(6.158). ■

For a specified terminal time t°, with the terminal state unrestricted, the necessary conditions (6.163)–(6.171) reduce to

$$\Gamma(t^\circ, u^\circ) = \mu^T \frac{\partial h(u^\circ)}{\partial u} \tag{6.172}$$

$$h(u^\circ) \geqq 0 \tag{6.173}$$

$$\mu^T h(u^\circ) = 0 \tag{6.174}$$

$$\mu \geqq 0 \tag{6.175}$$

$$\eta \geqslant 0. \tag{6.176}$$

Example 6.10 Constant velocity motion in a plane can be described by the state equations

$$\dot{x}_1 = \cos u \tag{6.177}$$

$$\dot{x}_2 = \sin u, \tag{6.178}$$

where the control u is the counterclockwise angle from the positive x_1-axis to the velocity vector. Suppose that the objectives are to transfer the system to the target $x_1 = 0$ in minimum time while staying as close to the x_1-axis as possible. More specifically, the costs are

$$G_1 = \int_0^t dt$$

$$G_2 = \int_0^t x_2^2\, dt,$$

with the target conditions

$$\theta_1(\cdot) = x_1 \leqq 0$$

$$\theta_2(\cdot) = -x_1 \leqq 0.$$

The state perturbation equations (6.159) are

$$\dot{\Lambda}_1 = -\sin u \tag{6.179}$$

$$\dot{\Lambda}_2 = \cos u \tag{6.180}$$

and the cost perturbation equation (6.161) is

$$\dot{\Gamma}=2\eta_2 x_2 \Lambda_2. \tag{6.181}$$

Integrating the state equations (6.177) and (6.178) with the initial state $x(0)$ yields

$$x_1=t\cos u+x_1(0) \tag{6.182}$$

$$x_2=t\sin u+x_2(0). \tag{6.183}$$

Integrating (6.179) and (6.180) with initial conditions $\Lambda_1(0)=\Lambda_2(0)=0$ yields

$$\Lambda_1=-t\sin u \tag{6.184}$$

$$\Lambda_2=t\cos u. \tag{6.185}$$

Substituting (6.183) and (6.185) in (6.181) and integrating, with the initial condition $\Gamma(0)=0$, yields

$$\Gamma=\eta_2 t^2\cos u\left[\tfrac{2}{3}t\sin u+x_2(0)\right]. \tag{6.186}$$

At a terminal time t the Pareto-minimal necessary conditions (6.163)–(6.171) become

$$\eta_1+\eta_2 x_2^2=(\pi_1-\pi_2)\cos u \tag{6.187}$$

$$\Gamma=(\pi_1-\pi_2)\Lambda_1 \tag{6.188}$$

$$x_1=0 \tag{6.189}$$

$$\eta_1\geqq 0$$

$$\eta_2\geqq 0$$

where

$$\eta_1+\eta_2>0.$$

Since the necessary conditions (6.186)–(6.188) are homogeneous in η and π, we may take

$$\eta_1=1-\eta_2 \tag{6.190}$$

with

$$0 \leqq \eta_2 \leqq 1. \tag{6.191}$$

We note, from (6.177), that $\cos u \neq 0$ for a trajectory that reaches the target $x_1 = 0$ starting from an initial point where $x_1(0) \neq 0$. Thus we may solve (6.187) for $\pi_1 - \pi_2$. Substituting this result into (6.188) and using (6.184), (6.186), and (6.190) yields

$$\eta_2 t^2 \cos u\left[\tfrac{2}{3} t \sin u + x_2(0)\right] = -\left\{1 - \eta_2 + \eta_2\left[t \sin u + x_2(0)\right]^2\right\} \tan u \tag{6.192}$$

where the final time, from (6.182) with $x_1 = 0$, is given by

$$t = -\frac{x_1(0)}{\cos u} > 0. \tag{6.193}$$

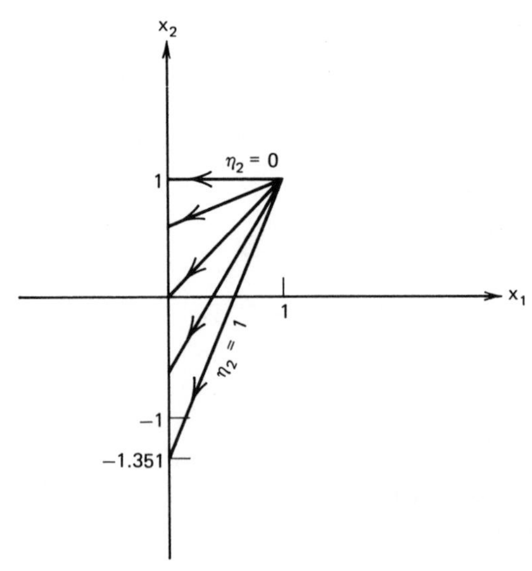

Figure 6.3. Pareto-minimal trajectories from (1, 1) for Example 6.10.

Substituting (6.193) into (6.192) and simplifying yields

$$\eta_2 x_1^2(0)\left[x_2(0)-\tfrac{2}{3}x_1(0)\tan u\right]=$$
$$-\left\{1-\eta_2+\eta_2\left[x_2(0)-x_1(0)\tan u\right]^2\right\}\sin u. \tag{6.194}$$

Thus for a given η_2, satisfying (6.191), and a given initial point $x_1(0), x_2(0)$, the corresponding Pareto-minimal control must satisfy (6.194), with the appropriate quadrant for u determined by (6.193). For the initial point $x_1(0)=x_2(0)=1$ we have

$$\eta_2\left(1-\tfrac{2}{3}\tan u\right)=-\left[1-\eta_2+\eta_2(1-\tan u)^2\right]\sin u \tag{6.195}$$

where $\cos u<0$. Figure 6.3 illustrates the Pareto-minimal trajectories from $(1,1)$ for $0\leqq\eta_2\leqq 1$. For $\eta_2=0$, $u=180°$. For $\eta_2=1$, $u=233.5°$.

6.6 EXERCISES

6.1 For linear systems the state perturbation matrix is directly related to the state transition matrix. Consider the linear system

$$\dot{x}=Ax+Bu$$

where A and B are constant $n\times n$ and $n\times s$ matrices, respectively. Given an initial point $x(0)$, the state at time t is given by

$$x(t)=\Phi(t)x(0)+\int_0^t \Phi(t-\tau)Bu\,d\tau$$

where $\Phi(t)$ is the $n\times n$ state transition matrix defined by the matrix initial value problem

$$\dot{\Phi}=A\Phi$$
$$\Phi(0)=I.$$

Show that for constant A the following relationship between the state perturbation matrix $\Lambda(t)$ and the state transition matrix $\Phi(t)=e^{At}$ is obtained:

$$A\Lambda(t)=\left[\Phi(t)-I\right]B.$$

6.2 Solve Example 6.4 with the target given by

$$-ax_1 - x_2 + b \leqq 0$$

where a and b are positive constants. Examine the specific case where $a=1$ and $b=2$.

6.3 Determine the constant control u that will minimize the time to drive the system

$$\dot{x}_1 = \cos u$$

$$\dot{x}_2 = \sin u$$

from the origin to the parabola $x_2 - x_1^2 + b = 0$. Use Theorem 6.2 with Λ and Γ determined directly from the definitions.

6.4 Determine the constant control u that will minimize the time to drive the system

$$\dot{x}_1 = x_2$$

$$\dot{x}_2 = u$$

from the origin to the line $x_2 = -x_1 + 1$ with $-1 \leqq u \leqq 1$.

6.5 Consider a simple dynamical system with the dynamics of the state given by

$$\dot{x}_1 = \cos u$$

$$\dot{x}_2 = \sin u.$$

Determine a constant control that will minimize the cost

$$G = 2\int_0^{t_f} [(x_1 - 1)\cos u + x_2 \sin u]\, dt$$

when the system starts at $x_1 = 0$, $x_2 = 3$ and terminates on the target specified by

$$\theta_1 = -x_1 \leqq 0$$

$$\theta_2 = x_1 - 2 \leqq 0$$

$$\theta_3 = x_2 - 1 \leqq 0$$

$$\theta_4 = -1 - x_2 \leqq 0.$$

6.6 Solve exercise 6.5 with the cost given by

$$G = -2\int_0^{t'} [(x_1 - 1)\cos u + x_2 \sin u]\, dt.$$

6.7 Make a plot of $u^* t^*$ versus $t^*/x(0)$ for $0 \leqq t^*/x(0) \leqq 100$ using (6.31). Compare this result on the same plot with Example 6.8 by finding numerical solutions for $u_1^* t^* = u_2^* t^*$ when $t^*/x_1(0) = t^*/x_2(0) = t^*/x(0)$. Examine the effect on yield. That is, plot $x(t^*)/x(0)$ and $x_1(t^*)/x(0) = x_2(t^*)/x(0)$ versus $t^*/x(0)$.

6.8 What would be the effect on yield if one of the genotypes of Example 6.8 were to use the optimal result of Example 6.2? Test your observations using the definition of a Nash equilibrium solution.

6.9 For certain biological problems the Nash concept may be further constrained (and simplified) by utilizing the "evolutionarily stable strategy" (ESS) concept introduced by Maynard Smith (1976). He defines an evolutionarily stable strategy for a biological system as one that, when common, would be fitter for a given mutant than any strategy generated by the mutation.

Consider a situation with many players, one of whom is the "mutant." Under ESS each player's control vector is identical. It also follows from the definition that the Nash criteria is satisfied by the mutant. Since any one of the players can be the mutant, some special structure is implied. In particular, the dynamics, cost criteria, and control constraint sets must be identical for every player.

Let the mutant player be identified by the scalar state x and his mutant strategy by the scalar control u. Let all the other players be characterized by a single scalar "state" z with each of the other players using a common scalar strategy v. Let the system dynamics be given by the scalar functions

$$\dot{x} = g_1(u, v, x, z)$$

$$\dot{z} = g_2(u, v, x, z)$$

with the common cost function given by

$$\dot{G} = \gamma(u, v, x, z).$$

Obtain necessary conditions for determining the ESS strategy for $u = v$ for a specified final time, final state unrestricted.

Appendix

ANSWERS TO SELECTED PROBLEMS

1.1 Global minimum at $y^*=-10$. $G(y^*)=-729$.

1.3 Minimum occurs when circle is tangent to $h(y)=0$ (i.e., radius vector perpendicular to $h(y)=0$). $y_1^*=\frac{3}{13}$, $y_2^*=\frac{9}{26}$, $G(y^*)=\frac{9}{52}$.

1.4 $y_1^*=5$, $y_2^*=\frac{1}{3}$, $G^*=-\frac{226}{9}$.

1.7 Rank $[A]=2$, rank $[B]=3$, rank $[C]=2$.

2.1 Use $G=-FN/(e^{-1}+N)$, $g=F-f(e^{-1})=0$. Condition (2.36) yields $\partial f/\partial e^{-1}=F/(e^{-1}+N)$.

2.2 8.58 percent.

2.3 At $(-1,0)$ and $(3,0)$, $(\partial G/\partial y)\,\delta y>0\ \forall y+\delta y\in Y$; however, $(-1,0)$ is a local maximum point. Examine first-order approximation for G in terms of δy [equation (1.15)]. Note that, as $\alpha\to 0, \delta y\to 0$ and the difference $G(y^*+\alpha\hat{\delta y})-G(y^*)$ is determined by $R(\alpha)$.

2.6 Beware! Since every point of Y is abnormal (and irregular), Theorem 2.1 may or may not hold.

2.8 $\cos\beta/V_s=\cos\alpha/V_w$ where $\beta=$heading angle in sand and $\alpha=$heading angle in water.

2.10 Yes, since the conditions for the theorem of Weierstrass are satisfied. Minimum candidate at $(1,1)$. Maximum candidates at $(1,1)$, $(2,\frac{1}{2})$, and $(\frac{1}{2},2)$. The second-order constraint qualification is not satisfied at noninternal candidates. Global maximum of $G(\cdot)$ may be determined by direct evaluation at the candidate points.

2.11 Candidates at $(0,0)$, $(0,8)$, $(\sqrt{\frac{31}{8}},\frac{1}{4})$, and $(-\sqrt{\frac{31}{8}},\frac{1}{4})$.

2.13 The diet is as follows: Eat every rabbit and, if $f_m/x_r t_r t_m>(f_r/t_r-f_m/t_m)$, then eat every mouse; if $f_m/x_r t_r t_m<(f_r/t_r-f_m/t_m)$, then eat no mice.

2.14 Candidate at $(3, \frac{39}{11}, \frac{26}{11})$.

2.15 Candidates at $(-2,0)$ and $(0,8)$.

3.1 Pareto-minimal candidate set $=\{y\in E^2 \mid y_1=y_2,\ 0\leqq y_1\leqq 1\}$.

3.2 Pareto-minimal candidate set $=\{y\in E^1 \mid y=0$ or $1\leqq y\leqq 3$ or $4\leqq y \leqq 10\}$.

3.3 Pareto-minimal candidate set $=\{y\in E^2 \mid y_2\geqq 5,\ y_1\leqq 5\}$.

3.4 Pareto-minimal candidate set $=\{y\in E^2 \mid y_2=0, 0\leqq y_1\leqq 1\}\cup\{y\in E^2 \mid 0 \leqq y_2\leqq \frac{1}{2},\ y_1=1+y_2\}$.

3.6 Pareto-minimal candidate set $=\{u\in E^2 \mid u_1+u_2=B,\ 0\leqq u_2\leqq u_1\}$

4.1 Local minimum at $(15,5)$.

4.7 Global maximum at $u=2$, $x=0$. Note that $g(\cdot)$ does not determine x as a function of u at the global maximum.

5.1 Pareto-minimal candidate set $=\{u\in E^2 \mid u_1=0,\ 0\leqq u_2\leqq 1\}$. Nash candidate $=(0,1)$. Min-max candidate for Player $1=(0,1)$. Min-max candidate for Player $2=(1,1)$. Bargaining set $=\{u\in E^2 \mid u_1=0,\ \frac{3}{5}\leqq u_2 \leqq 1\}$.

5.2 Pareto-minimal candidate set $=\{u\in E^2 \mid u_1=u_2,\ 0\leqq u_1\leqq \frac{1}{2}\}$. Nash candidate $=(0,\frac{1}{2})$. Min-max candidate for Player $1=(0,1)$. Min-max candidate for Player $2=(0,\frac{1}{2})$. Bargaining set $=\{u\in E^2 \mid u_1=u_2, (\sqrt{\frac{1}{2}}-\frac{1}{2})/\sqrt{2}\leqq u_1\leqq \frac{1}{2}\}$.

5.3 Pareto-minimal candidate set $=\{u\in E^2 \mid u_1=0,\ 0\leqq u_2\leqq 1\}\cup\{u\in E^2 \mid -1\leqq u_1\leqq 1,\ u_2=\frac{1}{2}\}\cup\{u\in E^2 \mid u_1=1,\ \frac{1}{2}\leqq u_2\leqq 1\}\cup\{u\in E^2 \mid u_1=-1, \frac{1}{2}\leqq u_2\leqq 1\}$. Nash candidate $=(0,1)$. Min-max candidate for Player $1=(0,1)\cup(0,0)\cup(0,-1)$. Min-max candidate for Player $2=(0,1)$. Bargaining set $=\{u\in E^2 \mid \sqrt{\frac{1}{2}}\leqq u_1\leqq \sqrt{\frac{3}{4}},\ u_2=\frac{1}{2}\}\cup \{u\in E^2 \mid -\sqrt{\frac{3}{4}}\leqq u_1\leqq -\sqrt{\frac{1}{2}},\ u_2=\frac{1}{2}\}$.

5.7 Pareto-minimal candidate set $=\{u\in E^2 \mid u_1=0,\ 0\leqq u_2\leqq 1\}\cup\{u\in E^2 \mid u_1=0,\ 2\leqq u_2<\infty\}\cup\{u\in E^2 \mid u_1=1,\ 0\leqq u_2<\infty\}$. Nash candidate $=$ min-max candidates $=(0,\infty)$. If the Utility moves first, it will use $u_2=1.999...$ and the Customer will respond with $u_1=1$. If the Customer moves first, he will use $u_1=0$ and the Utility will respond with $u_2=\infty$. From the viewpoint of an arbitrator (Public Utility Commission) it is noted the Customer will buy all the electricity the Utility can produce at any price up to K_1.

5.8 Species 2 should invade Species 1's territory slightly less than $\frac{1}{2}$ the time. Species 1 should spend no time repelling the invasion. The problem is worked using the Stackelburg solution concept with Species 2 moving first.

5.9 There is a point on the Pareto-minimal set for which both countries have the same cost. However, this point is not on the bargaining set. The "closest" point on the bargaining set to the Pareto-minimal equal cost point is given by the intersection of the Pareto-minimal set with a line of constant min-max cost for the first country.

5.13 Pareto-minimal candidate set $=\{(u,v)|v=2/3u-u/6,\ u\in[-2,-1]\cup[1,2]\}\cup\{(3,2),(-3,2)\}$. Nash candidates $=(u,v)\in\{(2,\frac{1}{4}),(-2,-\frac{1}{4}),(3,2),(-3,2),(-\frac{1}{4},-2)\}$. Min-max candidates for Player $1=(u,v)\in\{(1,\frac{1}{2}),(-1,-\frac{1}{2}),(\frac{1}{4},2)\}$. Min-max candidates for Player $2=\{u,v)\in\{(2,0),(-2,0),(0,-2)\}$.

6.5 $u=-71.565°$.

6.6 $u=-56.310°,\ -63.435°,\ -75.964°$.

REFERENCES

Blaquière, A., Gerard, F., and Leitmann, G., *Quantitative and Qualitative Games*, Academic, New York, 1969.

Bliss, G. A., *Lectures on the Calculus of Variations*, University of Chicago, Press, Chicago, Ill., 1946.

Boltyanskii, V. G., *Mathematical Methods of Optimal Control*, Holt, Reinhart and Winston, New York, 1971.

Bryson, A. E. and Ho, Y. C., *Applied Optimal Control*, revised printing, Wiley, New York, 1975.

Cesari, L., "Existence Theorems for Pareto Problems in Optimization," in *Calculus of Variations and Control Theory*, D. L. Russell, Ed., Pub. No. 36, Mathematics Research Center, University of Wisconsin-Madison, Academic, New York, 1976.

Coddington, E. A. and Levinson, N., *Theory of Ordinary Differential Equations*, McGraw-Hill, New York, 1955.

DaCunha, N. O. and Polak, E., "Constrained Minimization Under Vector-Valued Criteria in Finite Dimensional Spaces," *J. Math. Anal. Appl.*, Vol. 19, No. 1, 1967.

Farkas, J. I., "Uber die Theorie einfachen Ungleichungen," *J. Reine Angew. Math.*, Vol. 124, 1902.

Fiacco, A. V. and McCormick, G. P., *Nonlinear Programming: Sequential Unconstrained Minimization Techniques*, Wiley, New York, 1968.

Goh, B. S., *Management and Analysis of Biological Populations*, Elsevier, Amsterdam, 1980.

Grantham, W. J., "A Controllability Minimum Principle," Ph.D. dissertation, Aerospace and Mechanical Engineering, University of Arizona, Tucson, Az., 1973.

Grantham, W. J. and Vincent, T. L., "A Controllability Minimum Principle," *J. Optim. Theory Appl.*, Vol. 17, Nos. 1 and 2, 1975.

Hadley, G., *Linear Programming*, Addison-Wesley, Reading, Mass., 1962.

Hancock, H., *Theory of Maxima and Minima*, Dover, New York, 1960.

Hestenes, M.R., *Calculus of Variations and Optimal Control Theory*, Wiley, New York, 1966.

Holtzman, J. M. and Halkin, H., "Directional Convexity and the Maximum Principle for Discrete Systems," *SIAM J. Contr.*, Vol. 4, No. 2, 1966.

Isaacs, R., *Differential Games*, Wiley, New York, 1965.

Karush, W., "Minima of Functions of Several Variables with Inequalities as Side Conditions," Master's Thesis, Department of Mathematics, University of Chicago, Chicago, Ill., Dec. 1939.

Kuhn, H. W., "Nonlinear Programming: A Historical View," in *Nonlinear Programming, SIAM-AMS Proc.*, Vol. 9, American Mathematical Society, Providence, R.I., 1976.

Kuhn, H. W. and Tucker, A. W., "Nonlinear Programming," in J. Neyman (Ed.) *Proc. Second Berkeley Symposium on Mathematical Statistics and Probability*, University of California Press, Berkeley, 1951.

Lagrange, J. L., *Mécanique Analytique*, Paris, 1788.

Leitmann, G., *An Introduction to Optimal Control*, McGraw-Hill, New York, 1966.

Leitmann, G. and Schmitendorf, W. E., "Profit Maximization Through Advertising: A Nonzero Sum Differential Game Approach," *IEEE Trans. Automat. Contr.*, Vol. AC-23, No. 4, 1978.

Leitmann, G., Rocklin, S., and Vincent, T. L., "A Note on Control-Space Properties of Cooperative Games," *J. Optim. Theory Appl.*, Vol. 9, No. 6, 1972.

Lin, J. G., "Maximal Vectors and Multi-Objective Optimizations," in *Multicriteria Decision Making and Differential Games*, G. Leitmann, Ed., Plenum, New York, 1976a.

Lin, J. G., "Proper Equality Constraints and Maximization of Index Vectors," in *Multicriteria Decision Making and Differential Games*, G. Leitmann, Ed., Plenum, New York, 1976b.

Lin, J. G., "Proper Inequality Constraints and Maximization of Index Vectors," *J. Optim. Theory Appl.*, Vol. 21, No. 4, 1977.

Mangasarian, O. L., *Nonlinear Programming*, McGraw-Hill, New York, 1969.

May, R. M., Beddington, J. R., Clark, C. W., Holt, S. J., and Laws, R. M., "Management of Multispecies Fisheries," *Science*, Vol. 205, No. 4403, 1979.

Maynard Smith, J., "Evolution and the Theory of Games," *Amer. Sci.*, Vol. 64, 1976.

Miele, A., "Lagrange Multipliers and Quasi-Steady Flight Mechanics," *J. Aerosp. Sci.*, Vol. 26: 592–598, September 1959.

Nash, J. F., "Non-Cooperative Games," *Ann. Math.*, Vol. 54, No. 2, 1951.

Nering, E. D., *Linear Algebra and Matrix Theory*, Wiley, New York, 1963.

Pareto, V., *Cours d'Économie Politique*, Rouge, Lausanne, Switzerland, 1896.

Pontryagin, L. S., Boltyanskii, V. G., Gamkrelidze, R. V., and Mishchenko, E. F., *The Mathematical Theory of Optimal Processes*, Wiley, New York, 1962.

Pulliam, H. R., "On the Theory of Optimal Diets," *Am. Nat.*, Vol. 108, No. 959, 1974.

Rockafellar, C. T., *Convex Analysis*, Princeton University Press, Princeton, N.J., 1970.

Salukvadze, M. E., *Vector-Valued Optimization Problems in Control Theory*, Academic, New York, 1979.

Schmitendorf, W. E., "Cooperative Games and Vector-Valued Criteria Problems," *IEEE Trans. Automat. Contr.*, Vol. AC-18, No. 2, 1972.

Simaan, M. and Cruz, J. B., Jr., "On the Stackelberg Strategy in Nonzero-Sum Games," *J. Optim. Theory Appl.*, Vol. 11, No. 5, 1973.

Stadler, W., "A Survey of Multicriteria Optimization or the Vector Maximum Problem," *J. Optim. Theory Appl.*, Vol. 29, No. 1, 1979.

Starr, A. W. and Ho, Y. C., "Nonzero-Sum Differential Games," *J. Optim. Theory Appl.*, Vol. 3, No. 3, 1967.

Starr, A. W. and Ho, Y. C., "Further Properties of Non-Zero Sum Differential Games," *J. Optim. Theory Appl.*, Vol. 3, No. 4, 1969.

Takayama, A., *Mathematical Economics*, Dryden, Hinsdale, Ill., 1974.

Vincent, T. L., "Environmental Adaptation by Annual Plants (An Optimal Control/Games Viewpoint)," in *Lecture Notes in Control and Information Sciences*, Vol. 3, P. Hagedorn, Ed., Springer-Verlag, Berlin, 1977.

Vincent, T. L., "Game Theoretic Predictions of Genetic Shifts in a Plant Breeding Model," in *Proc. 21st Midwest Symposium on Circuits and Systems*, Aug 14–15, 1978a.

Vincent, T. L., "Yield for Annual Plants as an Adaptive Response," *Rocky Mountain J. Math.*, Vol. 9, No. 1, 1978b.

Vincent, T. L. and Brusch, R. G., "Optimal Endpoints," *J. Optim. Theory Appl.*, Vol. 6, No. 4, 1970.

Vincent, T. L. and Cliff, E. M., "Maximum-Minimum Sufficiency and Lagrange Multipliers," *AIAA J.*, Vol. 8, No. 1, 1970.

Vincent, T. L. and Goh, B. S., "Terminality, Normality, and Transversality Conditions," *J. Optim. Theory Appl.*, Vol. 9, No. 1, 1972.

Vincent, T. L. and Leitmann, G., "Control-Space Properties of Cooperative Games," *J. Optim. Theory Appl.*, Vol. 6, No. 2, 1970.

Vincent, T. L. and Mason, J. D., "Disconnected Optimal Trajectories," *J. Optim. Theory Appl.*, Vol. 3, No. 4, 1969.

Von Neumann, J. and Morgenstern, O., *Theory of Games and Economic Behavior*, Princeton University Press, Princeton, N.J., 1944.

Von Stackelberg, H., *The Theory of the Market Economy*, Oxford University Press, Oxford, England, 1952.

Yu, P. L., "Cone Convexity, Cone Extreme Points, and Nondominated Solutions in Decision Problems with Multiobjectives," in *Multicriteria Decision Making and Differential Games*, G. Leitmann, Ed., Plenum, New York, 1976.

Yu, P. L. and Leitmann, G., "Compromise Solutions, Domination Structures, and Salukvadze's Solution," in *Multicriteria Decision Making and Differential Games*, G. Leitmann, Ed., Plenum, New York, 1976.

Zadeh, L. H., "Optimality and Non-Scalar-Valued Performance Criteria," *IEEE Trans. Automat. Contr.*, Vol. AC-8, No. 1, 1963.

Zermelo, E., "Über das Navigationsproblem bei Ruhender Oder Veränderlicher Windverteilung," *Z. Angew. Math. Mech.*, Vol. 11, No. 2, 1931.

INDEX